SpringerBriefs in Computer Science

T0215139

More information about this series at http://www.springer.com/series/10028

Jian Zhang · Zhiqiang Zhang
Feifei Ma

Automatic Generation
of Combinatorial Test Data

 Springer

Jian Zhang
Zhiqiang Zhang
Feifei Ma
Institute of Software
Chinese Academy of Sciences
Beijing
China

ISSN 2191-5768   ISSN 2191-5776   (electronic)
ISBN 978-3-662-43428-4   ISBN 978-3-662-43429-1   (eBook)
DOI 10.1007/978-3-662-43429-1

Library of Congress Control Number: 2014948067

Springer Heidelberg New York Dordrecht London

Printed on acid-free paper

Springer is part of Springer Science+Business Media (www.springer.com)

Foreword

When software is designed using a large number of components, and is then executed in an environment that itself contains a large number of components, faults often arise from unexpected interactions among the components. How can we test these complex systems of interacting components? In particular, how can we find test suites that can reveal the most likely faults? The combinatorial explosion in the number of possible tests must be avoided; in principle, it can be avoided if faults arise from interactions among a few components. Yet generating small test suites remains a very challenging problem.

Some research concentrates on the underlying combinatorial problems and neglects the needs of the software tester; other research addresses testing without exploiting combinatorial information effectively. This book is valuable because it encompasses both the combinatorial theory and the testing practice. It formulates the construction of test suites as a combinatorial problem, clearly explaining the connection with testing needs. After introducing important relationships with studies in combinatorics, it focusses on algorithmic methods for the construction of test suites. The blending of software design, algorithms, and combinatorial mathematics provides an excellent introduction to this important topic.

Phoenix, AZ, USA, June 2014   Charles J. Colbourn

Preface

Testing is an important part of the software development life cycle. Many testing techniques have been proposed and used, to increase the quality and reliability of software and systems. Among them, combinatorial testing is a kind of black-box testing method that is gaining increasing popularity. It is particularly useful to reveal faults that are caused by the interaction of a few components/parameters of the system.

For many kinds of testing techniques, a challenging problem is to generate a small set of test cases which achieves certain coverage criterion. This is also true for combinatorial testing. Over the past 30 years or so, many different kinds of algorithms have been proposed. Some of them were designed specifically to solve the problem, while others are based on ideas from evolutionary computation and constraint processing communities.

This book intends to review the state of the art of automatic data generation methods for combinatorial testing. The first chapter introduces the basic concepts and notations that will be used later. It also gives a brief overview of the applications of combinatorial testing. The next five chapters describe various kinds of methods for generating input data for combinatorial testing. Each of these chapters can be read almost independently. Chapter 7 gives a list of tools that can be readily used. In the last chapter, we briefly discuss several other problems that are closely related to test data generation.

The book can be read by graduate students and researchers who work in the area of combinatorial testing (or testing in general). It should also be interesting to those who work in discrete mathematics, constraint solving, evolutionary computation, and other related fields. Practitioners may find the book useful when selecting a suitable technique and tool.

Due to the limit on the size of the book, we could not explain every algorithm in detail. We may also miss some important works in this field. You are welcome to contact us if you have any comments and suggestions.

Our research on software testing is partially supported by the 973 Program of China (grant No. 2014CB340701) and the National Science Foundation of China (NSFC grant No. 61100064). We are indebted to Charles Colbourn, Jacek

Czerkwonka, Lijun Ji, Changhai Nie, Zhilei Ren, Ziyuan Wang, Huayao Wu, Hantao Zhang, and Lie Zhu who read earlier drafts of this book and made many good suggestions. We would like to thank Jun Yan for his assistance during the preparation of this book. Thanks to Jeff Lei and Linbin Yu for answering our questions on the IPO algorithms. The editors at Springer, Celine Chang and Jane Li, are always helpful. We are grateful to them for their constructive suggestions and patience. In addition, the first author would like to thank Alfred Hofmann for his encouragement. Finally, we are very grateful to our families for their support.

Beijing, June 2014   Jian Zhang
                                                                              Zhiqiang Zhang
                                                                                 Feifei Ma

Contents

Chapter 1
Introduction to Combinatorial Testing

Abstract This chapter gives an introduction to combinatorial testing, in particular, an overview of its applications. Some basic concepts, including Latin squares, orthogonal arrays, and covering arrays, are described in detail, with examples. Variants of covering arrays, such as covering arrays with constraints, are also described.

1.1 Software Testing

Testing is an essential activity in software development. To ensure the correctness of programs, or software quality in general, we usually test the programs (software products) in one way or another.

There are many types of testing methods [31]. For example, we may distinguish between white-box testing and black-box testing, depending on the availability of source code. We also have various techniques for unit testing, integration testing, system testing, and regression testing, corresponding to different phases in the software development life cycle.

This book is about a specific kind of black-box testing technique—combinatorial testing. It can be used to test software and systems. The (software) system that is being tested is called the System Under Test (SUT). This book will focus on the test data generation problem, i.e., how to generate a suitable set of input data for combinatorial testing of the SUT.

1.2 Combinatorial Testing and Its Applications

In many cases (for example, during integration testing), the SUT has a number of components or parameters, each of which may take some values. For instance, in the system, we may consider the following components:

© The Author(s) 2014

J. Zhang et al., *Automatic Generation of Combinatorial Test Data*,
SpringerBriefs in Computer Science, DOI 10.1007/978-3-662-43429-1_1

- desktop, which can be manufactured by HP or Lenovo or Dell;
- operating system, which can be Linux or Microsoft Windows;
- web browser, which can be Microsoft IE or Firefox;
- database management system, which can be Oracle or MySQL;
- $\cdots$

For another example, suppose we are performing unit testing, and the function to be tested computes the product of two complex numbers: $(a_1 + b_1 i) \times (a_2 + b_2 i)$. So there are four parameters: a_1, b_1, a_2, b_2. For each parameter, the tester can select several representative values (like *Negative, zero, Positive*).

To test such a system, we may try all combinations of values and see if there is any error. But the cost might be too high, because the number of test cases increases exponentially when the number of parameters increases. Fortunately, experiences indicate that many systems exhibit erroneous behaviors due to the combination of just a few parameters' values. Thus, it is not necessary to check the combination of *all* parameters.

Wallace and Kuhn [41] analyzed software-related failures of some medical devices that led to recalls by the manufacturers. They found that many flaws could have been detected by testing all *pairs* of parameter settings. The applications are typically small to medium sized embedded systems.

Kuhn et al. [26] analyzed 329 error reports generated during the development and integration testing of a large distributed system developed at NASA Goddard Space Flight Center. The application is a data management system that gathers and receives large quantities of scientific data. They found that the number of conditions required to trigger a failure is at most six.

The above empirical studies show that software failures in many domains were caused by combinations of relatively few conditions. This motivates the use of combinatorial testing (CT), whose goal is to cover as many combinations of parameters or conditions as possible, using a relatively small test suite.

During the past 20 years or so, CT has been applied in many areas.

The AETG system [9] has been used in Bellcore for screen testing, interoperability testing, and protocol conformance testing. (Screen testing means testing the software that checks the user's inputs for consistency and validity.)

Cohen et al. [10] describes the application of CT to 10 Unix commands: `basename`, `cb`, `comm`, `crypt`, `sleep`, `sort`, `touch`, `tty`, `uniq`, and `wc`. In particular, they discussed several different ways to model the `sort` command.

Dunietz et al. [16] applied CT to an operations support system, which supports the maintenance operations to the AT&T network.

Smith et al. [37] described how to validate the Remote Agent Experiment (RAX) of the Deep Space One (DS1) mission. The authors advocate a combination of scenario-based testing and model-based validation. They used orthogonal arrays when designing test suites. (For an introduction to orthogonal arrays, see Sect. 1.3.2).

Ryu et al. [34] applied CT to testing the data part of the Transmission Control Protocol (TCP). Some typical discrete values were selected for some significant data parameters, and a set of test cases were generated to achieve pairwise coverage. Several system errors were found.

Lott et al. [29] uses an example, which is a basic billing system. The system processes telephone call data, and has the following four call properties:

- Access (which can be Loop, ISDN, or PBX);
- Billing (which can be Caller, Collect or 800);
- Call type (which can be Local, Long Distance, or International);
- Status (which can be Success, Busy or Blocked).

Kuhn and Okum [25] reports an experiment with a module of the Traffic Collision Avoidance System (TCAS). TCAS is a well-known benchmark in software testing and analysis. In the CT model, there are 12 parameters, which include two 10-value parameters, one 4-value parameter, two 3-value parameters and seven 2-value parameters.

Cohen et al. [13] used a simplified mobile phone product line as an example. The product line is artificial, but its structure reflects portions of the Nokia 6000 series phones. The model includes five components (two 2-value components and three 3-value components): Display, Email viewer, Camera, video camera and video ringtones. There are some constraints like, a graphical email viewer requires color display.

Yuan et al. [45] use CT in the testing of graphical user interfaces (GUIs). Each test case is a sequence of events, and the events may be repeated. The goal is to generate a number of test cases, with certain coverage.

Justin Hunter [24] performed an empirical study involving 10 projects at six companies. The experiments show that "CT can detect hard-to-find software faults more efficiently than manual test case selection methods".

Kuhn et al. [22] described the testing of smart phone applications. They studied the resource configuration file for Android apps, and found dozens of options. In the CT model, there can be nine parameters (like TOUCHSCREEN, KEYBOARD, SCREENLAYOUT_SIZE, etc.).

Garvin et al. [18] use a software product line for a media player as an example.

Wang et al. [42] described a black-box testing approach to detecting buffer overflow vulnerabilities. They noted that, in many cases, the attacker can influence the behavior of a target system only by controlling the values of its external parameters (e.g., input parameters, configuration options, and environment variables). They adapted the CT techniques to generate a test suite, so as to steer program execution to reach some vulnerable point.

Zhang et al. [46] conducted a case study on applying combinatorial testing to audio players to see whether they can correctly recognize and display information described in ID3v2 tags of audio files. Two CT models have been built for the ID3v2 tags based on two different test goals. It turns out that in some situations, a few audio players cannot display audio information correctly, and they may even crash.

In the book [23] (Chap. 5, "Test Parameter Analysis", by Eduardo Miranda), several case studies are given:

- a flexible manufacturing system which involves a conveyor, a camera, a machine vision system and a robotic arm subsystem. It picks up metal sheets of different shapes, colors, and sizes, and puts them into appropriate bins.
- an audio amplifier which has two input jacks, two volume controls, two toggle switches, and one three-way selection switch.
- a password diagnoser for an online banking system, which verifies that the user passwords conform to good security practices. For example, the length of a password should be at least eight (characters); a password should have at least one upper-case character, one numeric character and one special character.

Calvagna et al. [5] described how to perform conformance testing of a Java Card static verifier. They designed and implemented a combinatorial interaction based testing approach to the validation of the verifier component of Java Card virtual machines (JCVM), so as to check the degree of conformance to the verifier's design specifications.

Garn and Simos [17] presented a new testing framework called ERIS, to test the Linux system call APIs. The input space of the Linux system calls is modeled in terms of CT.

Vilkomir and Amstutz [40] applied the combinatorial approaches to the testing of mobile apps. The main goal is to reveal device-specific faults, and reduce the costs of testing. This is achieved by an algorithm for optimal selection of mobile devices.

1.3 Related Combinatorial Designs

Given a model of the SUT (with some parameters and the corresponding values), how can we come up with a small test suite that covers many combinations of parameters? That is a challenging problem. Researchers and engineers have tried to borrow some ideas and knowledge from the design of experiments.

This section describes some combinatorial designs [15] (including Latin squares, orthogonal arrays, and covering arrays) that are commonly used in experimental design. During the past 30 years, they have also been applied to software testing. They serve as abstract representations of combinatorial test suites. One benefit of such an abstraction is that they can be studied mathematically and algorithmically.

1.3.1 Latin Squares

In the early 1980s, Robert Mandl [30] used orthogonal Latin squares in compiler testing. In particular, the method was used in designing the Ada Compiler Validation Capability test suite. Later, Williams and Probert [43] also used orthogonal Latin squares in the testing of network interfaces.

Fig. 1.1 Orthogonal latin
squares

1 2 3 0	3 2 1 0	13 22 31 00
3 0 1 2	2 3 0 1	32 03 10 21
2 1 0 3	0 1 2 3	20 11 02 33
0 3 2 1	1 0 3 2	01 30 23 12
LS_1	LS_2	Mj

Definition 1.1 A *Latin square* of order n is an $n \times n$ array in which each cell is an element of a set D (whose size is n), such that each element occurs exactly once in each row and column.

Definition 1.2 A pair of Latin squares of order n is said to be *orthogonal* if the n^2 ordered pairs of elements formed by juxtaposing the two squares are all distinct. Formally, two latin squares L_1 and L_2 are said to be *orthogonal* if for any row x, y, and any column z, w,

$$(L_1(x, z) = L_1(y, w) \text{ \& } L_2(x, z) = L_2(y, w)) \rightarrow (x = y \text{ \& } z = w).$$

Here & denotes conjunction (AND), $\rightarrow$ denotes implication (IMPLIES).

Example 1.1 The matrices LS_1 and LS_2 in Fig. 1.1 are two Latin squares. When they are juxtaposed, as shown by the matrix Mj, we can see that all ordered pairs of cells are distinct. So LS_1 and LS_2 are orthogonal Latin squares.

A set of mutually orthogonal Latin squares (MOLS) has the property that every pair of Latin squares in the set is orthogonal.

1.3.2 Orthogonal Arrays

In the 1980s, *Orthogonal Arrays* (OAs) [3, 20, 33] were also used in software testing. For example, in the late 1980s, Phadke and his colleagues at AT&T developed the tool Orthogonal Array Testing System (OATS), and used it to test PMX/StarMAIL, an electronic mail product of AT&T, which is based on local area networks [4]. The product should work with several types and versions of LAN software, operating systems, and personal computers. OAs have also been used by Smith et al. [37], to generate test suites for validating the RAX of the DS1 mission.

Definition 1.3 An *orthogonal array* (OA) of *run size N, factor number k, strength t* is an $N \times k$ matrix having the following properties:

(1) There are exactly s_i symbols in each column i, $1 \le i \le k$.
(2) In every $N \times t$ subarray, each ordered combination of symbols from the t columns appears equally often in the rows.

The OA defined above can be denoted by $OA(N, s_1 \cdot s_2 \cdots s_k, t)$. We call s_i the *level* of factor i.

Fig. 1.2 OA(9, 3^3, 2)

```
0 0 2
0 1 1
0 2 0
1 0 1
1 1 0
1 2 2
2 0 0
2 1 2
2 2 1
```

When some factors have the same level, we may combine them and make the notation simpler. For example, 2^3 is the same thing as $2 \cdot 2 \cdot 2$; it means the OA has 3 factors, each of which has two possible values.

Figure 1.2 shows a small example of OA whose strength is 2. We can see that, for any two columns in the array, we get an even distribution of all the pairwise combinations of values. Specifically, in any 9×2 subarray, each pair of values appears exactly once. So the array is an OA.

For more examples, visit the library of OAs maintained by Sloane: http://neilsloane.com/oadir/.

Orthogonal arrays can be regarded as natural generalizations of orthogonal Latin squares. We can build an OA from a set of mutually orthogonal Latin squares (See Sect. 2.1.5).

When an OA is used as a test suite, each row in the array corresponds to a test case, and each factor corresponds to a component or parameter of the SUT.

Beautiful as they are, OAs have some weaknesses when used as test suites. In 1987, Tatsumi et al. [38, 39] proposed to relax the requirement of a combinatorial test suite. They think that it is enough if each tuple appears at least once; it is not necessary that each tuple appears the same number of times. In the early 1990s, Sherwood [35] also found limitations of OAs when applied to software/system testing. For example, sometimes the required OA does not exist. Even if the OA exists, its size is often quite large. Thus, in software testing, we usually do not use OAs, except for some small SUTs.

1.3.3 Covering Arrays

In the last 20 years, most people use covering arrays (CAs) [36] for CT.

Definition 1.4 A $CA(N, d_1 \cdot d_2 \cdots d_k, t)$ of *strength t* is an $N \times k$ array having the following properties:

(1) There are exactly d_i symbols in each column i ($1 \leq i \leq k$).
(2) In every $N \times t$ subarray, each ordered combination of symbols from the t columns appears at least once.

Each column of the CA corresponds to a factor or parameter p_i ($1 \leq i \leq k$), and d_i is called the *level* of p_i.

For each parameter/factor, the elements of the domain are not specified explicitly in the definition. They can be some concrete values like {John, Mary, …} or {NegativeInt, Zero, PositiveInt}. They may also be denoted by some abstract values like $\{1, 2, 3, \dots\}$ or $\{0, 1, 2, \dots\}$.

Similarly to the case of OAs, some parameters in a CA can be combined when their levels are the same. If every parameter has the same number of valid values, the array is called a *fixed level covering array* or simply a *CA*; otherwise, it is called a *mixed level covering array* (MCA) [11]. In the former case, we may denote the array by $CA(N, d^k, t)$, where $d = d_1 = \cdots = d_k$.

In the literature, some authors put t immediately after N. They use the notation $CA(N; t, d_1 \cdot d_2 \cdots d_k)$. When all the parameters have the same level v, they use $CA(N; t, k, v)$ or $CA(N; t, v^k)$.

Note that an OA requires that each t-tuple appears exactly the same number of times. In contrast, a CA only requires that any t-tuple appears *at least once* in any subarray. Thus, an OA is also a CA.

Some examples of (mixed) CAs (with strength $t = 2$) are given in Figs. 1.3 and 1.4. CAs like those in Fig. 1.3 are called *binary covering arrays*, because each parameter can take values from a binary alphabet (usually $\{0, 1\}$).

For the MCA in Fig. 1.4, the first parameter has three possible values, each of the other parameters is binary (which means, the value is either 0 or 1). If we take any two columns, say the first column and the last column, we shall see that the subarray

$$
\begin{array}{cc}
0 & 0 \\
0 & 1 \\
1 & 1 \\
1 & 0 \\
2 & 1 \\
2 & 0
\end{array}
$$

contains all possible value combinations: $\langle 0, 0 \rangle$; $\langle 0, 1 \rangle$; $\langle 1, 0 \rangle$; $\langle 1, 1 \rangle$; $\langle 2, 0 \rangle$; $\langle 2, 1 \rangle$.

Fig. 1.3 Two instances of $CA(5, 2^4, 2)$

```
0 0 0 0      0 0 0 1
0 1 1 1      0 0 1 0
1 0 1 1      0 1 0 0
1 1 0 1      1 0 0 0
1 1 1 0      1 1 1 1
```

Fig. 1.4 $MCA(6, 3 \cdot 2^4, 2)$

```
0 0 0 0 0
0 1 1 1 1
1 0 0 1 1
1 1 1 0 0
2 0 1 0 1
2 1 0 1 0
```

Fig. 1.5 Some values of
CAN (for $t = 2$)

k	N
3	4
4	5
10	6
15	7
35	8
56	9
126	10
210	11

$d = 2$

k	N
4	9
5	11
7	12
9	13
10	14
20	15
21	16
28	17

$d = 3$

k	N
10	76
11	78
13	84
15	96
16	102
18	104
19	107
20	108

$d = 8$

k	N
13	153
14	155
15	158
16	161
17	171
18	177
19	178
20	186

$d = 11$

When a CA is used as a test suite, each row of the array represents a test case, and the ith column corresponds to the values of parameter p_i in the test cases. Sometimes, testing using a CA of strength t is called t-way or t-wise testing. When the strength is 2, the testing is also called *pairwise testing*. In such a test suite, for any two components/parameters of the SUT, every pair of values appears in at least one test case.

In the literature, people use $CAN(d_1 \cdot d_2 \cdots d_k, t)$ to denote the smallest number N for which $CA(N, d_1 \cdot d_2 \cdots d_k, t)$ exists. It is called the *covering array number*. When all the d_i's are equal to v, we denote the number by $CAN(v, k, t)$ or $CAN(t, k, v)$. Thus, $CAN(t, k, v) = min\{N \mid \exists\, CA(N; t, k, v)\}$.

Some specific CANs are available at a website maintained by Charlie Colbourn: http://www.public.asu.edu/~ccolbou/src/tabby/catable.html

Just to give the readers an impression of the sizes of test suites, we list some known bounds[1] in Fig. 1.5.

1.4 Covering Arrays for Testing

When applying CAs to software testing, we often feel it necessary to extend the concept in various ways.

1.4.1 Seeding

Sometimes, there are some important parameter combinations that must be tested, such as the default configurations of newly-deployed software. *Seeding* first introduced by [8] allows testers to explicitly specify some parameter combinations (called *seeds*) to be

---

[1] Obtained from Colbourn's website, April 26, 2014. Note that the values given here are just the current best known results. It is possible that we will find smaller values in the future, for some cases.

covered in the test suite. A seed can be specified by a set of parameter-value pairs, like $\{(p_{i_1}, v_{i_1}), (p_{i_2}, v_{i_2}), \ldots, (p_{i_l}, v_{i_l})\}$.

In the rest of this book, we use the term *target combinations* to denote parameter combinations, which need to be covered as specified by the coverage requirement (covering strength or seeds).

1.4.2 Variable Strength Covering Arrays and Tuple Density

For CAs, the strength t is quite important. The larger it is, the more complete the testing will be. However, as t becomes larger, the number of test cases may increase rapidly.

The original definition of CAs requires all t-way parameter combinations to be covered. However, there are many cases where some parameters interact more (or less) with each other than with other parameters. If we enforce a global strength, the covering strength needs to be set at the highest interaction level, which will greatly increase the number of test cases, and a lot of resources will be wasted on testing unimportant parameter combinations. Cohen et al. [11, 12] introduced the concept of *variable strength covering arrays* (VCA), which allows the tester to specify different covering strengths on different subsets of parameters.

Definition 1.5 A variable strength $t^+ = \{(P_1, t_1), (P_2, t_2), \ldots, (P_l, t_l))\}$ is a set of coverage requirements, where P_i is a set of parameters, and t_i is a covering strength on P_i, for $1 \leq i \leq l$. Each requirement (P_i, t_i) requires that all t_i-way value combinations of parameters in P_i be covered.

When we replace the (universal) strength t with a variable strength t^+, the CA will be called a *variable strength covering array* (VCA). Note that the previous definition of strength t can be represented by a variable strength $t^+ = \{(\{p_1, p_2, \ldots, p_k\}, t)\}$. Thus, the variable strength CA is a generalization of the (traditional) CA. On the other hand, if a variable strength CA meets the covering requirement t^+, then for each $(P_i, t_i) \in t^+$, the subarray of parameters in P_i is a CA of strength t_i.

Figure 1.6 is an example of variable strength CAs. It is a CA of strength 2, except that for the last 4 columns, the strength is 3.

$$
\begin{array}{c|cccc}
2 & 1 & 1 & 1 & 1 \\
0 & 1 & 0 & 0 & 0 \\
1 & 0 & 0 & 1 & 1 \\
2 & 0 & 1 & 0 & 0 \\
1 & 1 & 1 & 1 & 0 \\
0 & 0 & 1 & 0 & 1 \\
2 & 1 & 0 & 0 & 1 \\
0 & 0 & 0 & 1 & 0 \\
1 & 1 & 1 & 0 & 1 \\
2 & 1 & 0 & 1 & 1 \\
2 & 0 & 0 & 0 & 1 \\
2 & 0 & 1 & 1 & 1 \\
\end{array}
$$

Fig. 1.6 $VCA(12, 3^1 2^4, 2, CA(12, 2^4, 3))$

In the traditional definition of CAs, the strength t is usually a positive integer. Chen and Zhang [7] refined the concept and proposed a new metric for combinatorial test suites, called *tuple density*. The metric considers coverage of tuples with dimensions higher than t; and it can be a rational number such as 3.4.

Definition 1.6 For a CA of strength t, the *tuple density* is the sum of t and the percentage of the covered $(t + 1)$-tuples out of all possible $(t + 1)$-tuples.

The tuple density is an extension of the strength t. It can be used to distinguish one test suite from another, even if they have the same size and strength.

1.4.3 Covering Arrays with Constraints

An important concept in CT is *constraints*. In some applications, some parameters in the SUT model must conform to certain restrictions, or else the test case will become invalid. For example, we may impose the constraint that "Internet Explorer does not run on Linux".

Ignoring parameter constraints will lead to *coverage holes*, i.e., this will make some test cases to be invalid and cannot be executed. If a combination is only covered by invalid test cases, then the failure caused by this combination will not be detected, since no test case containing it will be executed.

Now we give the definition of constraints as follows:

Definition 1.7 A constraint $c(v_{i_1}, v_{i_2}, \ldots, v_{i_j})$ is a predicate on the assignments of parameters $p_{i_1}, p_{i_2}, \ldots, p_{i_j}$. A test case $t = (v_1, v_2, \ldots, v_k)$ satisfies constraint c if and only if the assignment of those parameters makes the predicate true.

To the engineer (user of a CT tool), an important question is: How can the constraints be specified? A straightforward way is to list all the forbidden tuples; but this might be expensive. A more natural way is to use logic expressions to specify the constraints. But we need to choose a suitable language of logic. In [14], the constraints are modeled as Boolean formulas defined over atomic propositions that encode parameter-value pairs.

With the presence of constraints, some combinations of values may violate some of the constraints and cannot be in any valid test cases. They are called *unsatisfiable*. Also, the coverage requirement will be modified, which requires all satisfiable target combinations be covered by the array. To check the satisfiability of constraints, we may use a constraint solver or a satisfiability (SAT) solver for the propositional logic. For more details about constraint solving and SAT, see Sect. 6.1.

Arcuri and Briand [2] compared CT with random testing. They analyzed the issue formally. They obtained several theorems describing the probability of random testing to detect interaction faults. Their results indicate that random testing may outperform CT in some cases, if we do not consider constraints among features/parameters. But in the presence of constraints, random testing can be worse than CT. Thus, they suggest doing more research in CT for the cases in which constraints are present.

Very recently, Yilmaz [44] distinguishes between system-wide constraints and test case-specific constraints, and introduces a new combinatorial object, called a *test case-aware covering array*.

1.4.4 Covering Arrays with Shielding Parameters

In CT, we usually assume that all parameters of the system under test are always effective. But in some real applications, there may exist some parameters that can disable other parameters in certain conditions. A classical example is that, when we select "-h" (the help option) for a command, the other options are ignored. As another example, when we configure the TCP/IP protocol on a Microsoft Windows computer, we often need to fill in such information as IP address, subnet mask and default gateway. However, if we choose to obtain IP addresses *automatically*, these parameters are disabled.

To deal with the above problems, Chen et al. [6] proposed the concept Mixed Covering Arrays with Shielding parameters (MCAS).

1.4.5 Sequence Covering Arrays

In the classical formulation of CT, we assume that the parameters are somewhat independent of each other, subject to some declarative constraints on the values of certain parameters. However, in certain applications, the SUT accepts input parameters in some order, and then produces output values. In other words, the input to the SUT is a sequence of parameter values. When applying CAs to the testing of GUIs, Yuan et al. [45] leverage a "stateless" abstraction of GUIs.

Recently, to deal with the above kind of applications, Kuhn et al. [21] proposed the so-called sequence CAs. Such an array can be used to test the interactions of input sequences.

1.4.6 An Example of Using Covering Arrays for Testing

Now we give an example of using CAs for CT. We slightly modify the web application testing example that we mentioned earlier. There are four parameters/components in the system, which are shown in Table 1.1.

Since some web browsers can only be installed on specific operating systems, we need to write the following constraints to exclude invalid combinations:

```
Browser =="Internet Explorer" -> OS == "Microsoft Windows"
Browser == "Safari" -> OS == "Mac OS X"
```

Table 1.1 Parameters of a web application model

Operating system	Web browser	Flash player	Cookies
Microsoft windows	Internet explorer	10	Enabled
Linux	Firefox	11	Disabled
Mac OS X	Google chrome	12	
	Safari	13	
		Not installed	

Table 1.2 2-Way covering array for a web application model

OS	Browser	Flash	Cookies
Mac OS X	Google chrome	Not installed	Disabled
Mac OS X	Firefox	11	Enabled
Microsoft windows	Firefox	13	Disabled
Microsoft windows	Google chrome	10	Enabled
Mac OS X	Safari	12	Disabled
Microsoft windows	Internet explorer	11	Disabled
Linux	Firefox	10	Disabled
Linux	Google chrome	13	Enabled
Microsoft windows	Internet explorer	Not installed	Enabled
Microsoft windows	Google chrome	12	Enabled
Mac OS X	Safari	13	Enabled
Mac OS X	Safari	10	Enabled
Linux	Firefox	12	Disabled
Linux	Google chrome	11	Disabled
Linux	Firefox	Not installed	Disabled
Mac OS X	Safari	Not installed	Disabled
Mac OS X	Safari	11	Disabled
Microsoft windows	Internet explorer	10	Disabled
Microsoft windows	Internet explorer	12	Disabled
Microsoft windows	Internet explorer	13	Disabled

Here the symbol "− >" stands for logical implication. A->B enforces that if A is true, then B must be true. We know that there are some versions of Internet Explorer and Safari for other operating systems, but these versions are all discontinued and are rarely used. So it is reasonable to exclude these rare combinations.

A 2-way CA for the web application with regard to the constraints is shown in Table 1.2.

1.5 Pointers to the Literature

For the past three decades, many papers have been published in journals and conference proceedings in the fields of software engineering, theoretical computer science, discrete mathematics, and computational intelligence. Started in 2012, the International Workshop on Combinatorial Testing (IWCT) is held annually. The proceedings of the first three workshops (2012–2014) are available as part of the proceedings of the IEEE International Conference on Software Testing, Verification and Validation Workshops (ICSTW).

There are also some good (survey) articles and books in the literature, which provide rich sources of information about CT.

Kuhn et al. published the first book on CT [23], with some emphasis on the practical aspects of the technology. In addition to the basic concepts of CT, it covers many related topics, including random test generation, test suite prioritization, assertion-based testing and fault localization. The last chapter of the book presents algorithms for constructing CAs; but only two algorithms are described in detail: AETG and IPOG.

Mathur wrote a comprehensive book on software testing [31]. One chapter in that book is devoted to testing based on combinatorial designs. Another recent book on software testing, written by Ammann and Offutt [1], also has a section on combination strategies.

Grindal et al. published a survey [19], which presents 16 different combination strategies. The paper explores some properties of combination strategies, including coverage criteria and theoretical bounds on the size of test suites. It also attempts to relate the various coverage criteria, via a kind of subsumption hierarchy.

A more recent survey of CT is written by Nie and Leung [32]. Lawrence et al. [27] give a survey of binary CAs, including theorems about the existence of such arrays, bounds on CANs, and methods of producing binary CAs.

1.6 Structure of This Book

In software testing, a challenging task is to find a small test suite for a given SUT, which meets certain coverage criteria. For CT, we are mostly concerned with finding a small CA with a certain strength. It is shown [28] that the problem of generating a minimum pairwise test set is NP-complete. Thus, the test data generation problem for CT is a challenging problem. This book will describe various methods for constructing small combinatorial test suites satisfying certain requirements.

As mentioned earlier, CT is closely related to combinatorial design theory, which is a part of discrete mathematics. Mathematicians have been working in this area for a long time. Many results have been obtained, which may be useful in constructing CAs and OAs. Some of these results will be presented in Chap. 2.

Instead of using mathematical methods, which are usually applicable to some specific cases, we can also use computational methods, i.e., computer algorithms. One of the first influential computational method is the greedy algorithm of the automatic efficient test case generator (AETG) [8, 9]. This algorithm extends the CA one row at a time. It will be described in detail in Chap. 3.

Another important greedy approach to CA generation is the In Parameter Order (IPO) algorithm [28]. It is different from AETG in that it expands the array incrementally, both horizontally and vertically. The IPO algorithm and its descendants will be described in Chap. 4.

The problem of generating small CAs can be regarded as a constraint solving and optimization problem. In addition to problem-specific algorithms like AETG and IPO, we may also use many general-purpose search methods from artificial intelligence, operations research and evolutionary computation communities. Chapter 5 is devoted to the application of evolutionary algorithms (like genetic algorithms, simulated annealing, etc.) and tabu search to the generation of CAs. In Chap. 6, we shall describe backtracking search algorithms for finding CAs and OAs.

In Chap. 7, we list some available test generation tools and some well-known benchmarks. The reader can try these tools on the benchmarks, or use the tools to solve your own problems. Finally, in Chap. 8, we briefly touch some topics that are related to combinatorial test generation. They include, how to construct a CT model for the SUT, how to select a subset of test cases from an existing test suite, how to generate special-purpose test cases in the process of debugging, and so on.

References

1. Ammann, P., Offutt, J.: Introduction to Software Testing. Cambridge University Press, Cambridge (2008)
2. Arcuri, A., Briand, L.: Formal analysis of the probability of interaction fault detection using random testing. IEEE Trans. Softw. Eng. **38**(5), 1088–1099 (2012)
3. Bose, R.C., Bush, K.A.: Orthogonal arrays of strength two and three. Ann. Math. Stat. **23**(4), 508–524 (1952)
4. Brownlie, R., Prowse, J., Phadke, M.S.: Robust testing of AT&T PMX/StarMAIL using OATS. AT&T Tech. J. **71**(3), 41–47 (1992)
5. Calvagna, A., Fornaia, A., Tramontana, E.: Combinatorial interaction testing of a Java Card static verifier. In: Proceedings of the IEEE International Conference on Software Testing, Verification and Validation Workshops (ICSTW'14), pp. 84–87 (2014)
6. Chen, B., Yan, J., Zhang, J.: Combinatorial testing with shielding parameters. In: Proceedings of the 17th Asia Pacific Software Engineering Conference (APSEC), pp. 280–289 (2010)
7. Chen, B., Zhang, J.: Tuple density: a new metric for combinatorial test suites. In: Proceedings of the 33rd International Conference on Software Engineering (ICSE), pp. 876–879 (2011)
8. Cohen, D.M., Dalal, S.R., Fredman, M.L., Patton, G.C.: The AETG system: An approach to testing based on combinatorial design. IEEE Trans. Softw. Eng. **23**(7), 437–444 (1997)
9. Cohen, D.M., Dalal, S.R., Kajla, A., Patton, G.C.: The Automatic Efficient Test Generator (AETG) system. In: Proceedings of the 5th International Symposium on Software Reliability Engineering (ISSRE), pp. 303–309 (1994)
10. Cohen, D.M., Dalal, S.R., Parelius, J., Patton, G.C.: The combinatorial design approach to automatic test generation. IEEE Softw. **13**(5), 83–89 (1996)
11. Cohen, M. B., Gibbons, P. B., Mugridge, W. B., Colbourn, C. J.: Constructing test suites for interaction testing. In: Proceedings of the 25th International Conference on Software Engineering (ICSE), pp. 38–48 (2003)

12. Cohen, M.B., Gibbons, P.B., Mugridge, W.B., Colbourn, C.J., Collofello, J.S.: A variable strength interaction testing of components. In: Proceedings of the 27th Annual International Computer Software and Applications Conference (COMPSAC), pp. 413–418 (2003)
13. Cohen, M.B., Dwyer, M.B., Shi, J.: Interaction testing of highly-configurable systems in the presence of constraints. In: Proceedings of the International Symposium on Software Testing and Analysis (ISSTA), pp. 129–139 (2007)
14. Cohen, M.B., Dwyer, M.B., Shi, J.: Constructing interaction test suites for highly-configurable systems in the presence of constraints: A greedy approach. IEEE Trans. Softw. Eng. **34**(5), 633–650 (2008)
15. Colbourn, C.J., Dinitz, J.H. (eds.): Handbook of Combinatorial Designs, 2nd edn. Chapman & Hall/CRC, Boca Raton (2006)
16. Dunietz, I.S., Ehrlich, W.K., Szablak, B.D., Mallows, C.L., Iannino, A.: Applying design of experiments to software testing: Experience report. In: Proceedings of the 19th International Conference on Software Engineering (ICSE), pp. 205–215 (1997)
17. Garn, B., Simos, D.E.: Eris: A tool for combinatorial testing of the Linux system call interface. In: Proceedings of the IEEE International Conference on Software Testing, Verification and Validation Workshops (ICSTW' 14), pp. 58–67 (2014)
18. Garvin, B., Cohen, M., Dwyer, M.: Evaluating improvements to a meta-heuristic search for constrained interaction testing. Empir. Softw. Eng. **16**(1), 61–102 (2011)
19. Grindal, M., Offutt, J., Andler, S.F.: Combination testing strategies: A survey. Softw. Test. Verification Reliab. **15**(3), 167–199 (2005)
20. Hedayat, A.S., Sloane, N.J.A., Stufken, J.: Orthogonal Arrays: Theory and Applications. Springer, New York (1999)
21. Kuhn, D.R., Higdon, J.M., Lawrence, J., Kacker, R., Lei, Y.: Combinatorial methods for event sequence testing. In: Proceedings of the Fifth International Conference on Software Testing, Verification and Validation (ICST), pp. 601–609 (2012)
22. Kuhn, D.R., Kacker, R., Lei, Y.: Practical Combinatorial Testing. NIST Special Publication 800–142, Gaithersburg (2010)
23. Kuhn, D.R., Kacker, R., Lei, Y.: Introduction to Combinatorial Testing. Chapman & Hall / CRC, Boca Raton (2013)
24. Kuhn, R., Kacker, R., Lei, Y., Hunter, J.: Combinatorial software testing. Computer **42**(8), 94–96 (2009)
25. Kuhn, D.R., Okum, V.: Pseudo-exhaustive testing for software. In: Proceedings of the 30th Annual IEEE/NASA Software Engineering Workshop (SEW'06), pp. 153–158 (2006)
26. Kuhn, D.R., Wallace, D.R., Gallo Jr, A.M.: Software fault interactions and implications for software testing. IEEE Trans. Softw. Eng. **30**(6), 418–421 (2004)
27. Lawrence, J., Kacker, R.N., Lei, Y., Kuhn, D.R., Forbes, M.: A survey of binary covering arrays. Electron. J. Comb. **18**(1), 1–30 (2011)
28. Lei, Y. and Tai, K.C.: In-parameter-order: A test generation strategy for pair-wise testing. In: Proceeings of the 3rd IEEE International Symposium on High-Assurance Systems Engineering (HASE), pp. 254–261 (1998)
29. Lott, C., Jain, A., Dalal, S.: Modeling requirements for combinatorial software testing. In: Advances in Model-Based Testing (A-MOST 2005), ACM SIGSOFT Software Engineering Notes, **30**(4): 1–7 (2005)
30. Mandl, R.: Orthogonal Latin squares: An application of experiment design to compiler testing. Commun. ACM **28**(10), 1054–1058 (1985)
31. Mathur, A.P.: Foundations of Software Testing, 2nd edn. Pearson Education, Upper Saddle River (2013)
32. Nie, C., Leung, H.: A survey of combinatorial testing. ACM Comput. Surv. **43**(2), Article 11 (2011)
33. Rao, C.R. Factorial experiments derivable from combinatorial arrangements of arrays. J. Royal Stat. Soc., Supplement 9, pp. 128–139 (1947)
34. Ryu, J., Kim, M., Kang, S., Seol, S.: Interoperability test suite generation for the TCP data part using experimental design techniques. In: Proceedings of the International Conference on Testing Communicating Systems (TestCom), pp. 127–142 (2000)

35. Sherwood, G.: Efficient testing of factor combinations. In: Proceedings of the 3rd International Conference on Software Testing, Analysis and Review, (1994)

36. Sloane, N.J.A.: Covering arrays and intersecting codes. J. Comb. Des. **1**(1), 51–63 (1993)

37. Smith, B., Millar, W., Dunphy, J., Tung, Y.-W., Nayak, P., Gamble, E., Clark, M.: Validation and verification of the remote agent for spacecraft autonomy. In: Proceedings of the IEEE Aerospace Conference, vol. 1, pp. 449–468 (1999)

38. Tatsumi, K.: Test case design support system. In: Proceedings of International Conference on Quality Control (ICQC), pp. 615–620. Tokyo (1987)

39. Tatsumi, K., Watanabe, S., Takeuchi, Y., Shimokawa, H.: Conceptual support for test case design. In: Proceedings of the 11th International Computer Software and Applications Conf. (COMPSAC), pp. 285–290 (1987)

40. Vilkomir, S., Amstutz, B.: Using combinatorial approaches for testing mobile applications. In: Proceedings of the IEEE International Conference on Software Testing, Verification and Validation Workshops (ICSTW'14), pp. 78–83 (2014)

41. Wallace, D.R., Kuhn, D.R.: Failure modes in medical device software: An analysis of 15 years of recall data. Int. J. Reliab. Qual. Saf. Eng. **8**(4), 351–371 (2001)

42. Wang, W., Lei, Y., Liu, D., Kung, D., Csallner, C., Zhang, D., Kacker, R., Kuhn, R.: A combinatorial approach to detecting buffer overflow vulnerabilities. In: Proceedings of the IEEE/IFIP 41st International Conference on Dependable Systems & Networks, pp. 269–278 (2011)

43. Williams, A.W., Probert, R.L.: A practical strategy for testing pair-wise coverage of network interfaces. In: Proceedings of the Seventh International Symposium on Software Reliability Engineering (ISSRE'96), pp. 246–254 (1996)

44. Yilmaz, C.: Test case-aware combinatorial interaction testing. IEEE Trans. Softw. Eng. **39**(5), 684–706 (2013)

45. Yuan, X., Cohen, M. B., Memon, A. M.: Covering array sampling of input event sequences for automated GUI testing. In: Proceedings of the International Conference on Automated Software Engineering (ASE), pp. 405–408 (2007)

46. Zhang, Z., Liu, X., Zhang, J.: Combinatorial testing on ID3v2 tags of MP3 files. In: Proceedings of the IEEE Fifth International Conference on Software Testing, Verification and Validation (ICST), pp. 587–590 (2012)

Chapter 2
Mathematical Construction Methods

Abstract This chapter presents some representative mathematical methods that are commonly used in the construction of orthogonal arrays and covering arrays, as well as some bounds on the size of such arrays.

As we mentioned in Chap. 1, combinatorial testing is closely related to combinatorics, which is a branch of mathematics. For an introduction to combinatorial theory and combinatorial designs, see [3, 5]. Mathematicians have obtained many results in this field. Some of them can be used to construct combinatorial designs directly. Mathematical methods, in particular product or recursive constructions, can be employed to build large instances of orthogonal arrays and covering arrays. However, many of these methods are applicable only in specific cases.

2.1 Mathematical Methods for Constructing Orthogonal Arrays

As a combinatorial design with beautiful balancing property, the orthogonal array has long been the interest of mathematicians. There are many mathematical results about OA, either dealing with construction or proving its nonexistence given some parameters. For simplicity, here we just review a few ones that are easy to understand.

2.1.1 Juxaposition

Theorem 2.1 *If an $OA(N', s_1' \cdot s_2 \cdots s_k, t)$ and an $OA(N'', s_1'' \cdot s_2 \cdots s_k, t)$ both exist, and $\frac{N'}{s_1'} = \frac{N''}{s_1''}$, then an $OA(N' + N'', (s_1' + s_1'') \cdot s_2 \cdots s_k, t)$ exists.*

The proof of this theorem is trivial. Given an $OA(N', s_1' \cdot s_2 \cdots s_k, t)$ and an $OA(N'', s_1'' \cdot s_2 \cdots s_k, t)$, we just need to relabel the elements in the first column of one array, and put it underneath the other array. Obviously the resulting array is an $OA(N' + N'', (s_1' + s_1'') \cdot s_2 \cdots s_k, t)$.

© The Author(s) 2014

J. Zhang et al., *Automatic Generation of Combinatorial Test Data*,
SpringerBriefs in Computer Science, DOI 10.1007/978-3-662-43429-1_2

Fig. 2.1 Transform an
OA$(8, 4 \cdot 2^4, 2)$ to an
OA$(8, 2^6, 2)$ by splitting

$$
\begin{array}{c|cccc}
0 & 0\ 0\ 0\ 0 \\
0 & 1\ 1\ 1\ 1 \\
1 & 0\ 0\ 1\ 1 \\
1 & 1\ 1\ 0\ 0 \\
2 & 0\ 1\ 0\ 1 \\
2 & 1\ 0\ 1\ 0 \\
3 & 0\ 1\ 1\ 0 \\
3 & 1\ 0\ 0\ 1
\end{array}
\qquad
\begin{array}{cc|cccc}
0 & 0 & 0\ 0\ 0\ 0 \\
0 & 0 & 1\ 1\ 1\ 1 \\
0 & 1 & 0\ 0\ 1\ 1 \\
0 & 1 & 1\ 1\ 0\ 0 \\
1 & 0 & 0\ 1\ 0\ 1 \\
1 & 0 & 1\ 0\ 1\ 0 \\
1 & 1 & 0\ 1\ 1\ 0 \\
1 & 1 & 1\ 0\ 0\ 1
\end{array}
$$

2.1.2 Splitting

Theorem 2.2 *If an OA$(N, (p \times s) \cdot s_2 \cdots s_k, t)$ exists, then an OA$(N, p \cdot s \cdot s_2 \cdots s_k, t)$ also exists.*

For each element $i \in \{0, 1, \ldots, p \times s - 1\}$, let $v = i \bmod s$, $u = (i - v)/s$. Then split the first column of OA$(N, (p \times s) \cdot s_2 \cdots s_k, t)$ into two columns by replacing element i with two elements u and v. The resulting array is an OA$(N, p \cdot s \cdot s_2 \cdots s_k, t)$. For example, we can construct an OA$(8, 2^6, 2)$ from an OA$(8, 4 \cdot 2^4, 2)$ by splitting the first column, as illustrated in Fig. 2.1.

2.1.3 Hadamard Construction

An $n \times n$ $(-1, 1)$-matrix H_n is an Hadamard matrix if $H_n H_n^T = nI$.

Example 2.1 The matrix

$$
H_4 = \begin{pmatrix}
1 & 1 & 1 & 1 \\
1 & 1 & -1 & -1 \\
1 & -1 & 1 & -1 \\
1 & -1 & -1 & 1
\end{pmatrix}
$$

is an Hadamard matrix because:

$$
\begin{pmatrix}
1 & 1 & 1 & 1 \\
1 & 1 & -1 & -1 \\
1 & -1 & 1 & -1 \\
1 & -1 & -1 & 1
\end{pmatrix}
\times
\begin{pmatrix}
1 & 1 & 1 & 1 \\
1 & 1 & -1 & -1 \\
1 & -1 & 1 & -1 \\
1 & -1 & -1 & 1
\end{pmatrix}
=
\begin{pmatrix}
4 & 0 & 0 & 0 \\
0 & 4 & 0 & 0 \\
0 & 0 & 4 & 0 \\
0 & 0 & 0 & 4
\end{pmatrix}.
$$

Theorem 2.3 *When an Hadamard matrix of order $n \geq 4$ exists, an OA$(2n, 2^n, 3)$ and an OA$(8n, 2^{2n-4} \cdot 4^2, 3)$ exist.*

Due to space limitation, we just elaborate the first case. Suppose H_n is an Hadamard matrix of order n $(n \geq 4)$, then $\frac{H_n}{-H_n}$ is an OA$(2n, 2^n, 3)$. Figure 2.2 shows an OA$(8, 2^4, 3)$ constructed with the Hadamard matrix H_4 in the above example. By convention, we replace the element '-1' with '0'.

Fig. 2.2 Construct an
OA(8, 2^4, 3) from an
Hadamard matrix of order 4

$$\begin{pmatrix} 1 & 1 & 1 & 1 \\ 1 & 1 & -1 & -1 \\ 1 & -1 & 1 & -1 \\ 1 & -1 & -1 & 1 \\ \hline -1 & -1 & -1 & -1 \\ -1 & -1 & 1 & 1 \\ -1 & 1 & -1 & 1 \\ -1 & 1 & 1 & -1 \end{pmatrix}$$

```
1 1 1 1
1 1 0 0
1 0 1 0
1 0 0 1
0 0 0 0
0 0 1 1
0 1 0 1
0 1 1 0
```

2.1.4 Zero-Sum Construction

Theorem 2.4 *Suppose Z_s is the additive group of integers modulo s. For each of the s^t t-tuples over Z_s, form a row vector of length $t + 1$ by adjoining in the last column the negative of the sum of the elements in the first t columns. Use these vectors to form an $s^t \times (t + 1)$ array, and the array is an $OA(s^t, s^{t+1}, t)$.*

Example 2.2 Assume that we would like to construct an OA(16, 2^5, 4). Firstly, we enumerate all tuples of length 4 over the set $\{0, 1\}$, i.e., the 16 vectors ranging from $\langle 0, 0, 0, 0 \rangle$ to $\langle 1, 1, 1, 1 \rangle$. Secondly, for each vector we add an extra element in the end, so that the sum of all elements is divisible by 2. For instance, $\langle 0, 0, 0, 0 \rangle$ is extended to $\langle 0, 0, 0, 0, 0 \rangle$, and $\langle 0, 1, 1, 1 \rangle$ is extended to $\langle 0, 1, 1, 1, 1 \rangle$. Finally, all these vectors are used as row vectors to form the target array, as Fig. 2.3 illustrates.

2.1.5 Construction from Mutually Orthogonal Latin Squares

Theorem 2.5 *There exist k mutually orthogonal Latin squares of order n (k-MOLS(n)) if and only if there exists an $OA(n^2, n^{k+2}, 2)$.*

Fig. 2.3 OA(16, 2^5, 4)

```
0 0 0 0 0
0 0 0 1 1
0 0 1 0 1
0 0 1 1 0
0 1 0 0 1
0 1 0 1 0
0 1 1 0 0
0 1 1 1 1
1 0 0 0 1
1 0 0 1 0
1 0 1 0 0
1 0 1 1 1
1 1 0 0 0
1 1 0 1 1
1 1 1 0 1
1 1 1 1 0
```

Fig. 2.4 Construct an
OA(9, 3^4, 2) from
2-MOLS(3)

$$
\begin{array}{ccc}
 & 0\ 1\ 2 & 0\ 1\ 2 \\
 & 2\ 0\ 1 & 1\ 2\ 0 \\
 & 1\ 2\ 0 & 2\ 0\ 1
\end{array}
\qquad
\begin{array}{cccc}
0 & 0 & 0 & 0 \\
0 & 1 & 1 & 1 \\
0 & 2 & 2 & 2 \\
1 & 0 & 2 & 1 \\
1 & 1 & 0 & 2 \\
1 & 2 & 1 & 0 \\
2 & 0 & 1 & 2 \\
2 & 1 & 2 & 0 \\
2 & 2 & 0 & 1
\end{array}
$$

Suppose that $\{LS_1 \cdots LS_k\}$ is a set of k mutually orthogonal Latin squares of order n, and $LS_f(i, j)$ denotes the element in the ith row, jth column of LS_f. For each combination of i and j ($0 \leq i \leq n - 1$, $0 \leq j \leq n - 1$), we form a $(k + 2)$-tuple $\langle i, j, LS_1(i, j), \ldots, LS_k(i, j)\rangle$. Then we use these tuples as row vectors to form a matrix. Obviously the resulting matrix is an OA(n^2, n^{k+2}, 2).

Example 2.3 In Fig. 2.4, the matrix on the right is an OA(9, 3^4, 2) constructed from the 2-MOLS(3) on the left. The first column represents the row indices of 2-MOLS(3); the second column represents the column indices of 2-MOLS(3); the third column contains the elements in the first latin square, and the last column contains the elements in the second latin square.

2.2 Mathematical Methods for Constructing Covering Arrays

A covering array is optimal if it has the smallest possible number N of rows. As mentioned in Chap. 1, this smallest number is called the covering array number (CAN). Formally, CAN($d_1 \cdot d_2 \cdots d_k, t$) = min$\{N | \exists$CA($N, d_1 \cdot d_2 \cdots d_k, t)\}$. CAN is a vital attribute of covering arrays. It serves as the lower bound of the size of the test suite. CAN can always be obtained as the by-product of the construction of CAs.

2.2.1 Simple Constructions

2.2.1.1 Column-Collapsing

Given a CA($N, d_1 \cdots d_i \cdots d_k, t$), if we delete an arbitrary column i, we will get a CA($N, d_1 \cdots d_{i-1} \cdot d_{i+1} \cdots d_k, t$). So

$$\text{CAN}(d_1 \cdots d_{i-1} \cdot d_{i+1} \cdots d_k, t) \leq \text{CAN}(d_1 \cdots d_i \cdots d_k, t).$$

2.2.1.2 Symbol-Collapsing

Given a $CA(N, d_1 \cdots d_i \cdots d_k, t)$, if we replace a symbol v in the ith column with any symbol in this set $\{0, 1, \ldots, d_i - 1\}$ other than v itself, we will get a $CA(N, d_1 \cdots (d_i - 1) \cdots d_k, t)$. So

$$CAN(d_1 \cdots (d_i - 1) \cdots d_k, t) \leq CAN(d_1 \cdots d_i \cdots d_k, t).$$

2.2.1.3 Derivation

Given a $CA(N, d_1 \cdots d_i \cdots d_k, t)$, if we select a symbol v from the ith column, then extract the rows with a symbol v on the ith column, and finally delete the ith column, we will get a $CA(M, d_1 \cdots d_{i-1} \cdot d_{i+1} \cdots d_k, t - 1)$, where M is no less than the product of the $t - 1$ largest levels (excluding d_i). And we have

$$d_i \times CAN(d_1 \cdots d_{i-1} \cdot d_{i+1} \cdots d_k, t - 1) \leq CAN(d_1 \cdots d_i \cdots d_k, t).$$

2.2.1.4 Juxaposition

Given a $CA(N', d_1' \cdot d_2 \cdots d_k, t)$ and a $CA(N'', d_1'' \cdot d_2 \cdots d_k, t)$, we can construct a $CA(N' + N'', (d_1' + d_1'') \cdot d_2 \cdots d_k, t)$ by relabeling the symbols in the first column of one CA and putting it underneath the other.

In particular, we can construct a $CA(\ell N, \ell d_1 \cdot d_2 \cdots d_k, t)$ from a $CA(N, d_1 \cdot d_2 \cdots d_k, t)$, so we have

$$CAN(\ell d_1 \cdot d_2 \cdots d_k, t)) \leq \ell \cdot CAN(d_1 \cdot d_2 \cdots d_k, t).$$

For example, it is not difficult to find an instance of $CA(25, 5^2 4^1 3^2 2^7, 2)$. Then we can produce a $CA(100, 10^2 4^1 3^2 2^7, 2)$ by applying juxaposition twice. Obviously this result is optimal.

2.2.2 Recursive Constructions

For many combinatorial objects, we can use mathematical results to produce large objects from smaller ones. In this subsection, we briefly describe some of the results for covering arrays.

$$
\begin{array}{c}
\text{N rows} \\
\\
\\
\text{M rows} \\
\end{array}
\left|
\begin{array}{cccc}
a_{11} & a_{12} & \cdots & a_{1k} \\
a_{21} & a_{22} & \cdots & a_{2k} \\
 & & \vdots & \\
a_{N1} & a_{N2} & \cdots & a_{Nk} \\
b_{11} & b_{11} & \cdots & b_{11} \\
b_{21} & b_{21} & \cdots & b_{21} \\
 & & \vdots & \\
b_{M1} & b_{M1} & \cdots & b_{M1}
\end{array}
\right|
\begin{array}{cccc}
a_{11} & a_{12} & \cdots & a_{1k} \\
a_{21} & a_{22} & \cdots & a_{2k} \\
 & & \vdots & \\
a_{N1} & a_{N2} & \cdots & a_{Nk} \\
b_{12} & b_{12} & \cdots & b_{12} \\
b_{22} & b_{22} & \cdots & b_{22} \\
 & & \vdots & \\
b_{M2} & b_{M2} & \cdots & b_{M2}
\end{array}
\begin{array}{c}
\cdots \\
\cdots \\
\cdots \\
\cdots \\
\cdots \\
\cdots
\end{array}
\begin{array}{cccc}
a_{11} & a_{12} & \cdots & a_{1k} \\
a_{21} & a_{22} & \cdots & a_{2k} \\
 & & \vdots & \\
a_{N1} & a_{N2} & \cdots & a_{Nk} \\
b_{1\ell} & b_{1\ell} & \cdots & b_{1\ell} \\
b_{2\ell} & b_{2\ell} & \cdots & b_{2\ell} \\
 & & \vdots & \\
b_{M\ell} & b_{M\ell} & \cdots & b_{M\ell}
\end{array}
$$

Fig. 2.5 The product of $A \otimes B$ (Reprinted with permission from [9]. Copyright 2008, Elsevier Inc.)

Fig. 2.6 $\mathrm{CA}(N + (v - 1) \cdot M, v^{2k}, 3)$

$$
\begin{array}{ll}
A & A \\
B & B^{\pi^1} \\
B & B^{\pi^2} \\
 & \cdots \\
B & B^{\pi^{v-1}}
\end{array}
$$

2.2.2.1 Strength-2 Covering Arrays

For strength $t = 2$ (i.e., pairwise testing), Stevens and Mendelsohn proved the following theorem [8]:

Theorem 2.6 (Products of Strength-2 CAs) *If a* $\mathrm{CA}(N, v^k, 2)$ *and a* $\mathrm{CA}(M, v^\ell, 2)$ *both exist, then a* $\mathrm{CA}(N + M, v^{k\ell}, 2)$ *also exists.*

Let $A = (a_{ij})$ be $\mathrm{CA}(N, v^k, 2)$ and $B = (b_{ij})$ be $\mathrm{CA}(M, v^\ell, 2)$. Then the $(N + M) \times k\ell$ array $C = (c_{ij}) = A \otimes B$ in Fig. 2.5 is a $\mathrm{CA}(N + M, v^{k\ell}, 2)$.

Similarly, there are product methods for pairwise mixed covering arrays. For details, see [4]. The methods can be used recursively.

2.2.2.2 Strength-3 Covering Arrays

For covering arrays of strength 3, Chateauneuf and Kreher proved the following theorem [1]:

Theorem 2.7 *If a* $\mathrm{CA}(N, v^k, 3)$ *and a* $\mathrm{CA}(M, v^k, 2)$ *both exist, then a* $\mathrm{CA}(N + (v - 1) \cdot M, v^{2k}, 3)$ *also exists.*

Suppose A is a $\mathrm{CA}(N, v^k, 3)$, and B is a $\mathrm{CA}(M, v^k, 2)$. Let $\{\pi^i | 1 \le i \le v - 1\}$ be the cyclic group of permutations generated by $\pi = (0, 1, \ldots, v - 1)$. In other words, π^i is a bijection that maps symbol s to $(s + i) \bmod v$. Let B^{π^i} be the matrix obtained by applying the permutation π^i to B. Then a $\mathrm{CA}(N + (v - 1) \cdot M, v^{2k}, 3)$ can be constructed in the way illustrated in Fig. 2.6.

Fig. 2.7 Product
Construction for
$CA(13, 2^8, 3)$

```
0 0 0 0
1 1 1 1
1 0 0 1
1 0 1 0          0 0 0 0
0 1 0 1          0 1 1 1
1 1 0 0          1 0 1 1
0 0 1 1          1 1 0 1
0 1 1 0          1 1 1 0

A=CA(8,2⁴,3)     B=CA(5,2⁴,2)
```

```
0 0 0 0 | 0 0 0 0
1 1 1 1 | 1 1 1 1
1 0 0 1 | 1 0 0 1
1 0 1 0 | 1 0 1 0
0 1 0 1 | 0 1 0 1
1 1 0 0 | 1 1 0 0
0 0 1 1 | 0 0 1 1
0 1 1 0 | 0 1 1 0
0 0 0 0 | 1 1 1 1
0 1 1 1 | 1 0 0 0
1 0 1 1 | 0 1 0 0
1 1 0 1 | 0 0 1 0
1 1 1 0 | 0 0 0 1
```

Example 2.4 Suppose we are given two covering arrays, A is a $CA(8, 2^4, 3)$ and B is a $CA(5, 2^4, 2)$. Since $v = 2$, we have only one permutation π^1, which is $\pi^1(0) = 1$ and $\pi^1(1) = 0$. So B^{π^1} is actually obtained by permuting symbol '0' and '1' in B. We can form a $CA(13, 2^8, 3)$ with A, B and B^{π^1}, as shown in Fig. 2.7.

2.2.2.3 Covering Arrays of Arbitrary Strength

For covering arrays of arbitrary strength, we have the following theorem [1]:

Theorem 2.8 *If a* $CA(N, v^k, t)$ *and a* $CA(M, w^k, t)$ *both exist, then a* $CA(NM, (vw)^k, t)$ *also exists.*

Suppose A is a $CA(N, v^k, t)$, and B is a $CA(M, w^k, t)$. We construct a series of $N \times k$ matrices C_l with entries $C_l(i, j) =< A(i, j), B(l, j) >$, where $1 \leq i \leq N$, $1 \leq j \leq k$, and $1 \leq l \leq M$. Using these matrices, we form an $NM \times k$ matrix $C = [C_1, \ldots, C_M]^T$. Obviously C is a $CA(NM, (vw)^k, t)$.

2.2.3 Construction Based on Difference Covering Arrays

A difference covering array, or a $DCA(k, n; v)$ is an $n \times k$ array (d_{ij}) with entries from an Abelian group G of order v, such that for any two distinct columns l and h, the difference list

$$\delta_{l,h} = \{d_{1l} - d_{1h}, d_{2l} - d_{2h}, \ldots, d_{nl} - d_{nh}\}$$

contains every element of G at least once.

Example 2.5 The array in Fig. 2.8 is a $DCA(4, 3; 2)$ over Z_2 [10].

2.2.3.1 Strength-2 Covering Arrays

In [2], Colbourn showed that CAs of strength 2 can be easily constructed with DCAs.

Theorem 2.9 *If there exists a DCA$(k, n; v)$, then there exists a CA$(nv, v^k, 2)$.*

Given a DCA$(k, n; v)$ (d_{ij}) over the Abelian group G of order v. For each row vector

$$\langle d_{i1}, d_{i2}, \ldots, d_{ik} \rangle \quad (1 \le i \le n)$$

and each element $u \in G$, we construct a row vector

$$\langle (d_{i1} + u), (d_{i2} + u), \ldots, (d_{ik} + u) \rangle.$$

These $n \times v$ row vectors form a CA$(nv, v^k, 2)$.

Example 2.6 From the DCA$(4, 3; 2)$ in Fig. 2.8, we can construct a CA$(6, 2^4, 2)$ in Fig. 2.9 using this method. Since the DCA is over Z_2, each row vector in the DCA would produce two row vectors in the CA by adding '0' and '1' respectively.

2.2.3.2 Strength-3 Covering Arrays

In [6], Ji and Yin proposed two constructive methods to build covering arrays of strength 3 based on DCAs. For simplicity we only introduce the first one.

Theorem 2.10 *If there exists a DCA$(4, n; v)$, then there exists a CA$(nv^2, v^5, 3)$.*

Given a DCA$(4, n; v)$ (d_{ij}) over the Abelian group G of order v. For each row vector

$$\langle d_{i1}, d_{i2}, d_{i3}, d_{i4} \rangle \quad (1 \le i \le n),$$

we construct a series of row vectors

Fig. 2.8 DCA$(4, 3; 2)$

```
0 0 0 0
0 1 0 1
0 0 1 1
```

Fig. 2.9 Construct a CA$(6, 2^4, 2)$ from a DCA$(4, 3; 2)$

```
0 0 0 0
0 1 0 1
0 0 1 1
─────────
1 1 1 1
1 0 1 0
1 1 0 0
```

Fig. 2.10 Construct a
CA$(12, 2^5, 3)$ from a
DCA$(4, 3; 2)$

```
0 0 0 0 0
0 1 0 1 0
0 0 1 1 0
─────────
0 0 1 1 1
0 1 1 0 1
0 0 0 0 1
─────────
1 1 1 1 0
1 0 1 0 0
1 1 0 0 0
─────────
1 1 0 0 1
1 0 0 1 1
1 1 1 1 1
```

$$R(i, u, e) = \langle (d_{i1} + u), (d_{i2} + u), (d_{i3} + u + e), (d_{i4} + u + e), e \rangle,$$

where $u, e \in G$. These nv^2 row vectors form a CA$(nv^2, v^5, 3)$.

Example 2.7 From the DCA$(4, 3; 2)$ in Fig. 2.8, we can construct a CA$(12, 2^5, 3)$ in Fig. 2.10 using this method. Each row vector in the DCA would produce four row vectors in the CA since there are four value combinations of u and e.

There are also constructive methods to produce CAs of higher strength from DCAs. For example, CAs of strength 5 can be built with the methods in [7].

References

1. Chateauneuf, M., Kreher, D.: On the state of strength-three covering arrays. J. Comb. Des. **10**(4), 217–238 (2002)
2. Colbourn, C.J.: Combinatorial aspects of covering arrays. Le Matematiche (Catania) **58**, 121–167 (2004)
3. Colbourn, C.J., Dinitz, J.H. (eds.): Handbook of Combinatorial Designs, 2nd edn. Chapman & Hall / CRC, Boca Raton (2006)
4. Colbourn, C.J., Martirosyan, S.S., Mullen, G.L., Shasha, D., Sherwood, G.B., Yucas, J.L.: Products of mixed covering arrays of strength two. J. Comb. Des. **14**(2), 124–138 (2006)
5. Hall Jr, M.: Combinatorial Theory, 2nd edn. Wiley, New York (1998)
6. Ji, L., Yin, J.: Constructions of new orthogonal arrays and covering arrays of strength three. J. Comb. Theory Ser. A **117**, 236–247 (2010)
7. Ji, L., Li, Y., Yin, J.: Constructions of covering arrays of strength five. Des. Codes Cryptogr. **62**(2), 199–208 (2012)
8. Stevens, B., Mendelsohn, E.: New recursive methods for transversal covers. J. Comb. Des. **7**(3), 185–203 (1999)
9. Yan, J., Zhang, J.: A backtracking search tool for constructing combinatorial test suites. J. Syst. Softw. **81**(10), 1681–1693 (2008)
10. Yin, J.: Constructions of difference covering arrays. J. Comb. Theory Ser. A **104**(2), 327–339 (2003)

Chapter 3
One Test at a Time

Abstract This chapter presents the one-test-at-a-time algorithms for finding covering arrays. These algorithms generate the test suite in a greedy manner: test cases are generated one by one, until the coverage requirement is met, while each newly-generated test case covers as many uncovered target combinations as possible. Compared to other approaches, the one-test-at-a-time algorithms are more flexible to deal with various kinds of special test requirements, whoever the strategy brings some limitations on finding smaller test suites.

3.1 General Idea of the One-Test Strategy

The main idea of the strategy is straightforward: generate the test cases one by one (i.e., generate the covering array row by row) until the coverage requirement is met. During the process, each new test case covers as many uncovered target combinations as possible, so that the total number of test cases in the test suite can be minimized. The general framework of the one-test-at-a-time strategy can be abstracted as Algorithm 1.

Algorithm 1 The one-test-at-a-time Strategy

1: $test\_suite = \emptyset$
2: init($uncov\_target\_combs$);
3: **while** $uncov\_target\_combs \neq \emptyset$ **do**
4:   $new_test_case = $ gen_new_test_case()
5:   $test\_suite = test\_suite \cup \{new\_test\_case\}$
6:   update($uncov\_target\_combs, new\_test\_case$)
7: **end while**

At the beginning, the algorithm needs to initialize the set of all target combinations. In each iteration of the while-loop, it generates a test case that covers as many target combinations as possible. After a new test case is generated, the target combinations covered by the new test case are removed from the set. The while-loop terminates

© The Author(s) 2014
J. Zhang et al., *Automatic Generation of Combinatorial Test Data*,
SpringerBriefs in Computer Science, DOI 10.1007/978-3-662-43429-1_3

when all target combinations are covered. The main difference between algorithms based on this strategy resides in their different strategies to generate new test cases.

Here, we call a combination $\sigma = \{(p_{i_1}, v_{i_1}), (p_{i_2}, v_{i_2}), \ldots, (p_{i_l}, v_{i_l})\}$ to be *covered* by test case $\theta = (v'_1, v'_2, \ldots, v'_k)$ if and only if for $1 \leq j \leq l$, $v_{i_j} = v'_{i_j}$, i.e., the value of each parameter in σ is consistent with the corresponding value in θ. A combination σ is said to be covered by a test suite if and only if the test suite has at least one test case covering σ. If there is no test case in a test suite, which covers σ, the combination σ is said to be *uncovered*.

Note that with the presence of constraints, some target combinations may be *invalid*, i.e., no test case satisfying all the constraints can cover these combinations. As a consequence, the algorithm needs to give special consideration to the target combinations, since we need to know which target combinations are valid so that we can decide whether the current test suite meets the coverage requirement. Some algorithms check the validity of target combinations one by one during the initialization of the set of target combinations. But this may cost a lot of time when the number of target combinations is large.

3.2 Automatic Efficient Test Generator (AETG)

3.2.1 The AETG Algorithm

The one-test-at-a-time strategy was first used in AETG (Automatic Efficient Test Generator) [6]. The AETG algorithm is described as follows (suppose we are generating a covering array of strength t):

To generate each new test case, the algorithm first generates M candidate test cases (in the original AETG algorithm, $M = 50$), and selects the one covering the greatest number of uncovered target combinations as the new test case. For generating each candidate test case, AETG chooses the first parameter-value pair that appears the most frequently in the remaining target combinations (line 7 and 8). Then, it permutes the remaining parameters in random order (line 9), and greedily assigns values to them (line 10–13). For assigning each parameter, AETG chooses the value that makes the resulting partial test case cover the great number of uncovered target combinations.

To deal with constraints, as described in their patent [5], the original algorithm checks whether each candidate test case satisfies all the constraints after it is generated. If the new candidate test case does not satisfy all the constraints, it permutes the field labels to see whether the constraints are satisfied. If not, the algorithm abandons the candidate test case. If all the candidate test cases violate the constraints, the algorithm uses exhaustive search to find a valid test case.

We can see the constraint handling of the original AETG algorithm is inefficient, since when the number of constraints is large, most of the possible test cases will be invalid, so the candidate test cases generated by AETG are very likely to be invalid,

---

Algorithm 2 The AETG Algorithm

1: $test\_suite = \emptyset$
2: init($uncov\_target\_combs$);
3: **while** $uncov\_target\_combs \neq \emptyset$ **do**
4:     $candidates = \emptyset$
5:     **for** $m = 1$ to M **do**
6:       $candidate = empty\_test\_case$
7:       (p, v) = selectFirstParameterAndValue($uncov\_target\_combs$)
8:       Assign v to parameter p in $candidate$
9:       $paraOrder$ = permuteRemainingParameters()
10:       **for** each remaining parameter p in $paraOrder$ **do**
11:         v = selectValue(p, $uncov\_target\_combs$)
12:         Assign v to parameter p in $candidate$
13:       **end for**
14:       $candidates = candidates \cup \{candidate\}$.
15:     **end for**
16:     Choose the test case in $candidates$ covering the greatest number of uncovered target combinations as new_test_case.
17:     $test\_suite = test\_suite \cup \{new\_test\_case\}$
18:     update($uncov\_target\_combs$, new_test_case)
19: **end while**

---

thus the heuristic for maximizing the number of newly-covered target combinations will lose effectiveness. Cohen et al. [7, 9] proposed the AETG-SAT algorithm which is better in constraint handling. Before doing any assignment, the algorithm uses a SAT solver[1] to check the validity of the resulting partial test case. (If the assigned parameter is not related to any constraint, the validity check is not necessary). If the resulting partial test case is invalid, it tries with the next-best one, until a valid assignment is found. This method guarantees that at any moment, there must exists an assignment of the unassigned parameters that completes the partial test case into a valid test case.

3.2.2 AETG Example

Now we give an example to illustrate how AETG works. Suppose we are testing a web-based application on different client platforms to see whether it can work normally on different system configurations. Possible factors include the operating system (OS), web browser, whether Flash plugin is installed and the proxy configurations. For simplicity, we assume that each parameter has two values. The SUT model can be described as follows:

---

[1] For a brief introduction to SAT, see Sect. 6.1.

```
OS: Windows, Linux
Browser: IE, Firefox
Flash_Installed: Yes, No
Proxy: None, HTTP
```

Besides, we know the Internet Explorer (IE) browser can only be installed on Windows, so the following constraint is specified:

```
Browser=="IE" -> OS=="Windows"
```

For simplicity, we use p_1, p_2, p_3, p_4 to represent the four parameters, and use 1 and 2 to represent their values. The constraint is transformed into "$p_2 == 1 \rightarrow p_1 == 1$".

In this example, we use the AETG-SAT algorithm [7, 9] for illustration.

As the first step, the algorithm initializes the set of target combinations. Here, we use the vector form to represent parameter combinations. Suppose we have a combination $\sigma = \{(p_{i_1}, v_{i_1}), (p_{i_2}, v_{i_2}), \ldots, (p_{i_l}, v_{i_l})\}$. Its vector form is $(\hat{v}_1, \hat{v}_2, \ldots, \hat{v}_k)$, where for $1 \leq u \leq k$,

$$\hat{v}_u = \begin{cases} v_{i_j}, & \text{if there exists some } 1 \leq j \leq l, \text{ such that } u = i_j \\ -, & \text{otherwise} \end{cases}.$$

For the above SUT, the combination $\{(p_1, 1), (p_3, 2)\}$ can be represented by $(1, -, 2, -)$.

Now the set of all two-way parameter combinations is as follows:

$$\begin{array}{llll}
\{ & (1,1,-,-), & (1,2,-,-), & \rlap{\blacksquare\blacksquare\blacksquare} & (2,2,-,-), \\
& (1,-,1,-), & (1,-,2,-), & (2,-,1,-), & (2,-,2,-), \\
& (1,-,-,1), & (1,-,-,2), & (2,-,-,1), & (2,-,-,2), \\
& (-,1,1,-), & (-,1,2,-), & (-,2,1,-), & (-,2,2,-), \\
& (-,1,-,1), & (-,1,-,2), & (-,2,-,1), & (-,2,-,2), \\
& (-,-,1,1), & (-,-,1,2), & (-,-,2,1), & (-,-,2,2) \ \}.
\end{array}$$

It is easy to see that target combination $(2, 1, -, -)$ is invalid so it is excluded from the set of target combinations.

Now we show how the covering array is generated. For simplicity, the number of candidate test cases M is set as 2.

Table 3.1 Number of occurrences of parameter-value pairs (1st iteration)

p_1		p_2		p_3		p_4	
1	2	1	2	1	2	1	2
6	5	5	6	6	6	6	6

Iteration #1 For the first iteration, the number of occurrences of parameter-value pairs are shown in Table 3.1.

We choose p_1 and value 1, which appears the greatest number of times in uncovered target combinations, and is a valid parameter-value pair.

- To generate the first candidate test case, we choose p_2, p_3, p_4 as the order of the remaining parameters. According to the algorithm, a possible selection of values of these parameters is "1(valid), 1(valid), 1(valid)". So the resulting candidate is (1, 1, 1, 1).
- To generate the second candidate test case, we choose p_4, p_3, p_2 as the order of the remaining parameters. And a possible selection of values for this order is "2(valid), 1(valid), 1(valid)". So the resulting candidate is (1, 1, 1, 2).

Now we count the number of newly-covered target combinations. Each of the two candidate test cases covers 6 uncovered target combinations. We choose (1, 1, 1, 1) as the first test case.

Now we remove the target combinations covered by the first test case. So the set of uncovered target combinations becomes:

$$\{ \quad (1,2,\text{-},\text{-}), \quad (2,2,\text{-},\text{-}),$$
$$(1,\text{-},2,\text{-}), \quad (2,\text{-},1,\text{-}), \quad (2,\text{-},2,\text{-}),$$
$$(1,\text{-},\text{-},2), \quad (2,\text{-},\text{-},1), \quad (2,\text{-},\text{-},2),$$
$$(\text{-},1,2,\text{-}), \quad (\text{-},2,1,\text{-}), \quad (\text{-},2,2,\text{-}),$$
$$(\text{-},1,\text{-},2), \quad (\text{-},2,\text{-},1), \quad (\text{-},2,\text{-},2),$$
$$(\text{-},\text{-},1,2), \quad (\text{-},\text{-},2,1), \quad (\text{-},\text{-},2,2) \quad \}.$$

Iteration #2 For the second iteration, the number of occurrences of parameter-value pairs are shown in Table 3.2.

We choose p_2 and value 2, which now appears the greatest number of times in uncovered target combinations, and is a valid parameter-value pair.

- To generate the first candidate test case, we choose p_1, p_3, p_4 as the order of the remaining parameters, and a possible selection of values for this order is "1(valid), 2(valid), 2(valid)". The resulting candidate is (1, 2, 2, 2).
- To generate the second candidate test case, we choose p_3, p_4, p_1 as the order of the remaining parameters, and a possible selection of values for this order is "1(valid), 2(valid), 2(valid)". The resulting candidate is (2, 2, 1, 2).

Table 3.2 Number of occurrences of parameter-value pairs (2nd iteration)

p_1		p_2		p_3		p_4	
1	2	1	2	1	2	1	2
3	5	2	6	3	6	3	6

Table 3.3 Number of occurrences of parameter-value pairs (3rd iteration)

p_1		p_2		p_3		p_4	
1	2	1	2	1	2	1	2
0	5	2	3	3	3	3	3

Now we count the number of newly-covered target combinations. Both of the two candidate test cases cover 6 uncovered target combinations. We choose $(1, 2, 2, 2)$ as the second test case.

Now we remove the target combinations covered by the second test case. So the set of uncovered target combinations becomes:

$$\{ \quad \text{░░░} \quad \text{░░░} \quad \text{░░░} \quad (2,2,\text{-},\text{-}),$$
$$\text{░░░} \quad \text{░░░} \quad (2,\text{-},1,\text{-}), \quad (2,\text{-},2,\text{-}),$$
$$\text{░░░} \quad \text{░░░} \quad (2,\text{-},\text{-},1), \quad (2,\text{-},\text{-},2),$$
$$\text{░░░} \quad (\text{-},1,2,\text{-}), \quad (\text{-},2,1,\text{-}), \quad \text{░░░}$$
$$\text{░░░} \quad (\text{-},1,\text{-},2), \quad (\text{-},2,\text{-},1), \quad \text{░░░}$$
$$\text{░░░} \quad (\text{-},\text{-},1,2), \quad (\text{-},\text{-},2,1), \quad \text{░░░} \quad \}.$$

Iteration #3 For the third iteration, the number of occurrences of parameter-value pairs are shown in Table 3.3.

We choose p_1 and value 2, which now appears the greatest number of times in uncovered target combinations, and is a valid parameter-value pair.

- For generating the first candidate test case, we choose p_2, p_3, p_4 as the order of the remaining parameters, and a possible selection of values for this order is "2(valid), 1(valid), 1(valid)". The resulting test case is $(2, 2, 1, 1)$.
- For generating the second candidate test case, we choose p_3, p_2, p_4 as the order of the remaining parameters, and a possible selection of values for this order is "2(valid), 1(invalid, choose 2 instead), 1(valid)". The resulting test case is $(2, 2, 2, 1)$. Note that when selecting value for p_2, the first choice was 1, which makes the partial test case $(2, 1, 2, -)$ violates constraint "$p_2 == 1 \rightarrow p_1 == 1$", so the value was abandoned, and 2 was chosen.

Both of the two candidate test cases cover 5 uncovered target combinations. We choose $(2, 2, 1, 1)$ as the third test case.

Now we remove the target combinations covered by the third test case. So the set of uncovered target combinations now becomes:

$$\{ \quad \text{░░░} \quad \text{░░░} \quad \text{░░░} \quad \text{░░░},$$
$$\text{░░░} \quad \text{░░░} \quad \text{░░░}, \quad (2,\text{-},2,\text{-}),$$
$$\text{░░░} \quad \text{░░░} \quad \text{░░░}, \quad (2,\text{-},\text{-},2),$$
$$\text{░░░} \quad (\text{-},1,2,\text{-}), \quad \text{░░░}, \quad \text{░░░}$$
$$\text{░░░} \quad (\text{-},1,\text{-},2), \quad \text{░░░}, \quad \text{░░░}$$
$$\text{░░░} \quad (\text{-},\text{-},1,2), \quad (\text{-},\text{-},2,1), \quad \text{░░░} \quad \}.$$

Table 3.4 Number of occurrences of parameter-value pairs (4th iteration)

p_1		p_2		p_3		p_4	
1	2	1	2	1	2	1	2
0	2	2	0	1	3	1	3

Iteration #4 For the fourth iteration, the number of occurrences of parameter-value pairs are shown in Table 3.4.

We choose p_3 and value 2, which now appears the greatest number of times in uncovered target combinations, and is a valid parameter-value pair.

- To generate the first candidate test case, we choose p_2, p_1, p_4 as the order of the remaining parameters, and a possible selection of values for this order is "1(valid), 2(invalid, choose 1 instead), 2(valid)". The resulting test case is $(1, 1, 2, 2)$. Note that when selecting value for p_1, the first choice was 1, which makes the partial test case $(2, 1, 2, -)$ violates constraint "$p_2 == 1 \rightarrow p_1 == 1$", so the value was abandoned, and 1 was chosen.
- To generate the second candidate test case, we choose p_1, p_2, p_4 as the order of the remaining parameters, and a possible selection of values for this order is "2(valid), 1(invalid, choose 2 instead), 2(valid)". The resulting test case is $(2, 2, 2, 2)$. Note that when selecting value for p_2, the first choice was 1, which makes the partial test case $(2, 1, 2, -)$ violates constraint "$p_2 == 1 \rightarrow p_1 == 1$", so the value was abandoned, and 2 was chosen.

Both of the two candidate test cases cover two uncovered target combinations. We choose $(1, 1, 2, 2)$ as the fourth test case.

Now we remove the target combinations covered by the fourth test case. So the set of uncovered target combinations now becomes:

$$\{ \; \blacksquare \quad \blacksquare \quad \blacksquare \quad \blacksquare ,$$
$$\blacksquare \quad \blacksquare \quad \blacksquare , \quad (2,-,2,-),$$
$$\blacksquare \quad \blacksquare \quad \blacksquare , \quad (2,-,-,2),$$
$$\blacksquare \quad \blacksquare \quad \blacksquare , \quad \blacksquare$$
$$\blacksquare \quad \blacksquare \quad \blacksquare , \quad \blacksquare$$
$$\blacksquare \quad (-,-,1,2), \quad (-,-,2,1), \quad \blacksquare \; \}.$$

Iteration #5 The rest of the test generation process is described briefly. For the fifth iteration, p_1 and value 2 are selected (valid). And for parameter order p_3, p_4, p_2, values "2(valid), 2(valid), 1(invalid, choose 2 instead)" are selected, and candidate $(2, 2, 2, 2)$ is generated. Then for parameter order p_2, p_3, p_4, values "1(invalid, choose 2 instead), 2(valid), 2(valid)" are selected, and candidate $(2, 2, 2, 2)$ is generated. The two candidates are identical and cover 2 uncovered target combinations, and finally the first candidate is chosen. Then, the set of remaining target combinations becomes:

Table 3.5 Test suite generated by AETG

p_1	p_2	p_3	p_4
1	1	1	1
1	2	2	2
2	2	1	1
1	1	2	2
2	2	2	2
1	1	1	2
1	1	2	1

{ ▨▨▨▨ ▨▨▨▨ ▨▨▨▨ ▨▨▨▨,
  ▨▨▨▨ ▨▨▨▨ ▨▨▨▨, ▨▨▨▨
  ▨▨▨▨ ▨▨▨▨ ▨▨▨▨, ▨▨▨▨
  ▨▨▨▨ ▨▨▨▨ ▨▨▨▨, ▨▨▨▨
  ▨▨▨▨ ▨▨▨▨ ▨▨▨▨, ▨▨▨▨
  ▨▨▨▨ (-,-,1,2), (-,-,2,1), ▨▨▨▨ }.

Iteration #6 For the sixth iteration, p_3 and value 1 are selected (valid). And for parameter order p_1, p_2, p_4, values "1(valid), 1(valid), 2(valid)" is selected, and candidate (1, 1, 1, 2) is generated. Then for parameter order p_2, p_1, p_4, values "2(valid), 2(valid), 2(valid)" is selected, and candidate (2, 2, 1, 2) is generated. Both candidates covers 1 uncovered target combinations, and finally the first candidate is chosen. Then, the set of remaining target combinations becomes:

{ ▨▨▨▨ ▨▨▨▨ ▨▨▨▨ ▨▨▨▨,
  ▨▨▨▨ ▨▨▨▨ ▨▨▨▨, ▨▨▨▨
  ▨▨▨▨ ▨▨▨▨ ▨▨▨▨, ▨▨▨▨
  ▨▨▨▨ ▨▨▨▨ ▨▨▨▨, ▨▨▨▨
  ▨▨▨▨ ▨▨▨▨ ▨▨▨▨, ▨▨▨▨
  ▨▨▨▨ ▨▨▨▨ (-,-,2,1), ▨▨▨▨ }.

Similarly, for the seventh iteration, test case (1, 1, 2, 1) is generated, and all target combinations are covered. The complete test suite is (Table 3.5).

3.3 AETG Variants

There are many other algorithms based on the one-test-at-a-time strategy. They can be regarded as variants of AETG.

Bryce et al. [3] introduced a four-layer framework for AETG-like algorithms which construct (mixed) CAs one row at a time. Many algorithms can be instantiated from this framework. The layers are the following:

- Layer one—Repetitions. We may generate covering arrays a number of times, keeping the smallest one. This layer is meaningful only when some decision in the algorithm is made randomly.
- Layer two—Candidate rows. An algorithm may generate a number of candidate rows, choosing the best one to add to the covering array.
- Layer three—Factor ordering. Factors may be ordered either randomly, or by the number of levels associated with a factor, or by the number of uncovered pairs involving this factor and the fixed factors, etc.
- Layer four—Level selection. For a given factor f, we can select some level (value) for it using some level selection criterion. For example, the candidates can be ranked by the number of uncovered pairs involving this level (value) of the current factor and the fixed factors.

As we mentioned earlier, AETG-SAT [7, 9] is based on the AETG algorithm, and uses a SAT solver to check the validity of the resulting partial test case before any assignments are made. The algorithm handles constraints better than the original AETG algorithm.

Jacek Czerwonka's tool PICT [10] uses a similar strategy to greedily assign values to parameters. Like AETG-SAT, it checks constraint satisfiability when attempting to assign a value to an unassigned parameter. PICT translates the constraints into forbidden combinations, and computes other combinations that are forbidden as implied by the constraints. The satisfiability check is done by checking whether the partial test case contains any explicit or implicit forbidden combinations. It can also guarantee the assigned partial test case is always a part of a valid test case.

The TCG algorithm [11] uses a similar strategy with AETG, but the parameter assignment process is deterministic: (1) for adding each row, it generates only one test case instead of multiple candidates; (2) when assigning parameter values, it selects the least used value; (3) parameters are assigned in nonincreasing order of their domain sizes. The DDA algorithm [1, 2] uses a density-based approach to decide which parameter and value to choose. It is also deterministic.

The algorithm proposed by Calvagna and Gargantini [4] generates each test case by greedily selecting a set of consistent uncovered target combinations, and then using a model checker to assign the remaining parameter values. Constraint satisfiability is checked when attempting to add a new uncovered target combination to the selected set of target combinations. For each row, the algorithm also generates only one test case instead of multiple candidates.

3.4 Test Generation Using Constrained Optimization

What the AETG-like algorithms we described in the last section have in common is that when generating each test case, they all proceed in a greedy manner, which assign parameters in a certain order, and choose values for the parameters using some strategy to maximize the number of newly-covered target combinations.

It is easy to see that generating each new test case is an optimization process, i.e., finding a test case that maximizes the number of newly-covered target combinations. There are dozens of well-known optimization techniques, such as meta-heuristic search algorithms and branch-and-bound algorithms. Several one-test-at-a-time algorithms based on heuristic search and optimization techniques have been proposed. For more details, see Chap. 5.

The problem of generating a new test case can also be translated into logic-based optimization problem. In our CT test generation tool *Cascade* [12, 13], the problem is translated into a linear pseudo-Boolean optimization (PBO) problem. The linear pseudo-Boolean optimization problem is identical to the 0–1 integer programming problem, which has the following elements:

- A set of Boolean variables, each of which takes the value of 0 or 1.
- A set of linear pseudo-Boolean (PB) constraints specifying the restrictions the solutions should conform to.
- A pseudo-Boolean objective function specifying the goal to optimize.

Here, a PB constraint or a PB function is *linear* if it contains no multiplication of Boolean variables. An example of linear PBO problem is as follows:

$$\begin{cases} \min : 2 * x_1 + 3 * x_3 - 1 * x_5 \\ 1 * x_1 + 3 * x_4 + 1 * x_5 \geq 2 \\ 1 * x_2 + 2 * x_3 + 1 * x_4 = 1 \\ \dots \end{cases} \quad (3.1)$$

Now we translate the problem of generating each test case into a PBO problem in the following way:

- For each parameter p and one of its values v, we use a variable $A_{(p,v)}$, which is true if and only if parameter p takes value v.
- For each target combination σ, we use a variable B_σ, which is true if and only if σ is covered by the test case.
- For each parameter p, suppose its value domain is $\{v_1, v_2, \ldots, v_{s_i}\}$. The parameter can take exactly one value. So we add the following PB constraint:

$$A_{(p,v_1)} + A_{(p,v_2)} + \cdots + A_{(p,v_{s_i})} = 1 \quad (3.2)$$

- For each combination $\sigma = \{(p_{i_1}, v_{i_1}), (p_{i_2}, v_{i_2}), \ldots, (p_{i_l}, v_{i_l})\}$, it is covered if and only if for $1 \leq j \leq l$, parameter p_{i_j} takes value v_{i_j}. The following PB constraints are generated:

$$\begin{aligned} -B_\sigma + A_{(p_{i_1}, v_{i_1})} &\geq 0; \\ -B_\sigma + A_{(p_{i_2}, v_{i_2})} &\geq 0; \\ &\cdots \\ -B_\sigma + A_{(p_{i_l}, v_{i_l})} &\geq 0; \\ \sum_{1 \leq j \leq l} -A_{(p_{i_j}, v_{i_j})} + B_\sigma &\geq -l + 1; \end{aligned} \quad (3.3)$$

- To deal with a constraint, it is first translated into the form of forbidden combinations. For each forbidden combination $\sigma = \{(p_{i_1}, v_{i_1}), (p_{i_2}, v_{i_2}), \ldots, (p_{i_l}, v_{i_l})\}$, the following PB constraint is generated:

$$- A_{(p_{i_1}, v_{i_1})} - A_{(p_{i_1}, v_{i_2})} - \cdots - A_{(p_{i_1}, v_{i_l})} \geq -l + 1; \qquad (3.4)$$

- The objective function to minimize is the summation of the negation of all uncovered target combinations. Suppose the set of all uncovered target combinations is U, then the objective function is:

$$\min : \sum_{\sigma \in U} -B_{\sigma} \qquad (3.5)$$

After the PBO problem is generated, the algorithm calls an external PBO solver to solve the translated problem. The solution of the PBO problem can be easily translated back into a test case. Besides, the solver can find the optimal solution for the PBO problem, which corresponds to the test case that covers the greatest number of uncovered target combinations. However, this will cost a lot of solving time and may not contribute much to the reduction of the test suite size. Certain strategies could be applied to prematurely stop the solver to get a near-optimal solution in a relatively short period of time.

The validity of target combinations could be checked one by one during the initialization of the set of target combinations. This may cost a lot of time when the number of target combinations is large. But fortunately, there is another way to work around. The validity of each target combination does not need to be checked during initialization. The algorithm just continues trying to generate new test cases covering as many uncovered target combinations as possible. If in a certain iteration, no more target combinations can be covered by a valid test case, the test generation process terminates, and all the remaining uncovered target combinations are invalid.

3.4.1 Example of PBO-Based Algorithm

Again, we generate a test suite for the SUT model of the web application example. Suppose at a certain point of time, the generated test cases are as shown in Table 3.6.

Table 3.6 Partial test suite generated by PBO-based algorithm

p_1	p_2	p_3	p_4
1	2	2	1
1	1	1	2
2	2	2	2
2	2	1	1

The set of uncovered target combinations is:

{ ▨▨▨ ▨▨▨ (2,1,-,-) ▨▨▨,
▨▨▨ ▨▨▨ ▨▨▨, ▨▨▨
▨▨▨ ▨▨▨ ▨▨▨, ▨▨▨
▨▨▨ (-,1,2,-), ▨▨▨, ▨▨▨
(-,1,-,1), ▨▨▨ ▨▨▨, ▨▨▨
▨▨▨ ▨▨▨ ▨▨▨, ▨▨▨ }.

Note that we do not check the satisfiability of target combinations, so combination $(2, 1, -, -)$ is still in the set of uncovered target combinations. To generate the next test case, a PBO problem is generated:

$$
\left\{
\begin{array}{ll}
\min : B_{(2,1,-,-)} + B_{(-,1,2,-)} + B_{(-,1,-,1)} & \\
\quad A_{(p_1,1)} + A_{(p_1,2)} & = 1; \\
\quad A_{(p_2,1)} + A_{(p_2,2)} & = 1; \\
\quad A_{(p_3,1)} + A_{(p_3,2)} & = 1; \\
\quad A_{(p_4,1)} + A_{(p_4,2)} & = 1; \\
\quad -B_{(2,1,-,-)} + A_{(p_1,2)} & \geq 0; \\
\quad -B_{(2,1,-,-)} + A_{(p_2,1)} & \geq 0; \\
\quad -A_{(p_1,2)} - A_{(p_2,1)} + B_{(2,1,-,-)} & \geq -1; \\
\quad -B_{(-,1,2,-)} + A_{(p_2,1)} & \geq 0; \\
\quad -B_{(-,1,2,-)} + A_{(p_3,2)} & \geq 0; \\
\quad -A_{(p_2,1)} - A_{(p_3,2)} + B_{(-,1,2,-)} & \geq -1; \\
\quad -B_{(-,1,-,1)} + A_{(p_2,1)} & \geq 0; \\
\quad -B_{(-,1,-,1)} + A_{(p_4,1)} & \geq 0; \\
\quad -A_{(p_2,2)} - A_{(p_4,1)} + B_{(-,1,-,1)} & \geq -1; \\
\quad -A_{(p_1,2)} - A_{(p_2,1)} & \geq -1;
\end{array}
\right.
$$

Note that the last PB constraint is translated from constraint "$p_2 = 1 \rightarrow p_1 = 1$", which forbids combination $(2, 1, -, -)$.

The PBO problem is solved by the solver, and the solution is $(1, 1, 2, 1)$. We accept this test case as the fifth test case, and update the set of uncovered target combinations, which now becomes:

{ ▨▨▨ ▨▨▨ (2,1,-,-) ▨▨▨,
▨▨▨ ▨▨▨ ▨▨▨, ▨▨▨
▨▨▨ ▨▨▨ ▨▨▨, ▨▨▨
▨▨▨ ▨▨▨ ▨▨▨, ▨▨▨
▨▨▨ ▨▨▨ ▨▨▨, ▨▨▨
▨▨▨ ▨▨▨ ▨▨▨, ▨▨▨ }.

After this, a similar process is conducted, and the solution covers no new target combinations, so the test generation terminates and the remaining target combination $(2, 1, -, -)$ is unsatisfiable. The complete test suite is as shown in Table 3.7.

Table 3.7 Test suite
generated by PBO-based
algorithm

p_1	p_2	p_3	p_4
1	2	2	1
1	1	1	2
2	2	2	2
2	2	1	1
1	1	2	1

References

1. Bryce, R.C., Colbourn, C.J.: The density algorithm for pairwise interaction testing. Softw. Test. Verification Reliab. **17**(3), 159–182 (2007)
2. Bryce, R.C., Colbourn, C.J.: A density-based greedy algorithm for higher strength covering arrays. Softw. Test. Verification Reliab. **19**(1), 37–53 (2008)
3. Bryce, R.C., Colbourn, C.J., Cohen, M.B.: A framework of greedy methods for constructing interaction test suites. In: Proceedings of the International Conference on Software Engineering (ICSE), pp. 146–155 (2005)
4. Calvagna, A., Gargantini, A.: A Logic-Based Approach to Combinatorial Testing with Constraints. Tests and Proofs, pp. 66–83. Springer, Heidelberg (2008)
5. Cohen, D.M., Dalal, S.R., Fredman, M.L., Patton, G.C.: Method and system for automatically generating efficient test cases for systems having interacting elements. US Patent 5,542,043, 30 July 1996
6. Cohen, D.M., Dalal, S.R., Fredman, M.L., Patton, G.C.: The AETG system: an approach to testing based on combinatorial design. IEEE Trans. Softw. Eng. **23**(7), 437–444 (1997)
7. Cohen, M.B., Dwyer, M.B., Shi, J.: Interaction testing of highly-configurable systems in the presence of constraints. In: Proceedings of the 2007 International Symposium on Software Testing and Analysis, pp. 129–139 (2007)
8. Cohen, M.B., Dwyer, M.B., Shi, J.: Exploiting constraint solving history to construct interaction test suites. In: Proceedings of Testing: Academic and Industrial Conference Practice and Research Techniques-MUTATION (TAICPART-MUTATION 2007), pp. 121–132 (2007)
9. Cohen, M.B., Dwyer, M.B., Shi, J.: Constructing interaction test suites for highly-configurable systems in the presence of constraints: a greedy approach. IEEE Trans. Softw. Eng. **34**(5), 633–650 (2008)
10. Czerwonka, J.: Pairwise testing in the real world: practical extensions to test-case scenarios. In: Proceedings of the 24th Pacific Northwest Software Quality Conference (PNSQC'06), pp. 419–430 (2006)
11. Tung, Y.W., Aldiwan, W.S.: Automating test case generation for the new generation mission software system. In: Proceedings of the IEEE Aerospace Conference, vol. 1, pp. 431–443 (2000)
12. Zhao, Y., Zhang, Z., Yan, J., Zhang, J.: Cascade: a test generation tool for combinatorial testing. In: Proceedings of the IEEE Sixth International Conference on Software Testing, Verification and Validation Workshops (ICSTW'13), pp. 267–270 (2013)
13. Zhang, Z., Yan, J., Zhao, Y., Zhang, J.: Generating combinatorial test suite using combinatorial optimization. J. Syst. Softw., to appear (2014)

Chapter 4
The IPO Family

Abstract Aside from the one-test-at-a-time strategy, the in-parameter-order (IPO) is another greedy strategy for generating covering arrays. As the name says, the IPO strategy extends the covering array in parameter order. In each iteration, the algorithm extends a smaller covering array into a larger covering array with one more parameter. The extension process has two stages, which extends the covering array horizontally and vertically, respectively.

4.1 In-Parameter-Order

The in-parameter-order (IPO) strategy was first introduced by Lei and Tai [3], and was later modified and generalized as the in-parameter-order-general (IPOG) test generation framework [4]. Different from the one-test-at-a-time strategy, which extends the covering array row by row, the IPO strategy extends the covering array in parameter order: it starts from a small covering array of a subset of parameters, and gradually extends the small covering array into larger covering arrays. Suppose at a certain moment, the current covering array is a set of the first i parameters, the IPO strategy first extends the covering array by adding an additional column so that the resulting array with $i + 1$ columns covers as many target combinations as possible, and then it extends the array vertically by adding new rows to cover the remaining uncovered target combinations of the first $i + 1$ parameters. And finally, when the array is extended to k columns, it becomes a complete covering array for the k parameters.

The execution of the IPO algorithms can be illustrated as Fig. 4.1.

The original IPOG algorithm is shown in Algorithm 3, which is intended to be used to generate covering arrays for some fixed covering strength t.

The algorithm first initializes the test suite as all combinations of the first t parameters in line 2. Then, it enters a for-loop in line 3. In the ith iteration of the for-loop, the algorithm starts from a covering array of the first $i - 1$ parameters, and extends the covering array to a larger covering array of the first i parameters. The extension process has two stages: a horizontal extension and a vertical extension. The horizontal extension stage adds one column for the ith parameter, covering as many new target combinations as possible. After horizontal extension, the array may still

J. Zhang et al., *Automatic Generation of Combinatorial Test Data*,
SpringerBriefs in Computer Science, DOI 10.1007/978-3-662-43429-1_4

Fig. 4.1 IPO algorithms

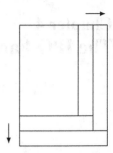

Algorithm 3 The IPOG Algorithm Framework

1: Denote the parameters in some order, as $p_1, p_2, \ldots p_k$
2: Set *test_suite* as all combinations of parameter values $p_1, p_2, \ldots, p_t$
3: **for** $i = t + 1$ to k **do**
4:     Let π be the set of t-way combinations of values involving parameter p_i and $t - 1$ parameters
       among the first $i - 1$ parameters
5:     // *horizontal extension for parameter* p_i
6:     **for** each test $\tau = (v_1, v_2, \ldots, v_{i-1})$ in *test_suite* **do**
7:         Choose a value v_i of p_i and replace τ with $\tau' = (v_1, v_2, \ldots, v_{i-1}, v_i)$ so that τ' covers the
           greatest number of target combinations in π
8:         Remove from π the target combinations covered by τ'
9:     **end for**
10:    // *vertical extension for parameter* p_i
11:    **for** each combination σ in π **do**
12:        **if** there exists a test that already covers σ **then**
13:            Remove σ from π
14:        **else**
15:            Change an existing test, if possible, or otherwise add a new test to cover σ and remove it
               from π
16:        **end if**
17:    **end for**
18: **end for**

---

have some target combinations uncovered, so the vertical extension stage is used to cover these target combinations by modifying existing test cases or adding new test cases. After the for-loop in line 3 terminates, the resulting array is a covering array of all the k parameters of strength t.

In the vertical extension stage, when adding a new test case to cover a target combination σ, only the parameters in σ are actually assigned their respective values in σ, other values are *don't-care values*, denoted by "$*$". When modifying a test case to cover σ, the algorithm first finds a test case that is consistent with σ, which means for each parameter in σ, its value in the test case is either the same with its value in σ, or a don't-care value.

4.2 IPOG-C for Handling Constraints

Yu et al. [10] proposed the IPOG-C algorithm to add constraint handling support to the IPOG framework. Several modifications are made:

- During the initialization stage in line 2, the covering array is initialized as all valid combinations of the first t parameters.
- When initializing the set of target combinations in line 4, the validity of each target combination should be checked. Only valid target combinations are stored in the set.
- In the horizontal extension stage, the chosen value for parameter p_i should make τ' be valid and make it cover the greatest number of uncovered target combinations.
- In the vertical extension stage, when modifying a test case to cover a target combination, the resulting test case must still be valid.

When checking the validity of a target combination or a partial test case, IPOG-C translates the problem into a constraint satisfaction problem (CSP),[1] and a constraint solver is used to check whether a feasible solution exists. The target combination or the partial test case is valid if and only if the resulting CSP has a solution.

The IPOG-C algorithm provides several optimizations to enhance the performance of constraint handling, including the following:

- The validity of target combinations can be postponed to the moment between horizontal extension and vertical extension. This is because every newly-covered target combinations in horizontal extension must be valid. So checking the validity of target combinations after horizontal extension can greatly reduce the number of target combinations that need to be checked, and the number of solver calls can also be reduced.
- When checking the satisfiability of a target combination, or checking the satisfiability of a partial test case after replacing don't-care values with concrete values, a constraint relation graph is built to find all relevant constraints that may be violated when some parameter values change. As a result, the number of constraints need to be checked in CSP could be reduced, and the solving time could be reduced too.
- During the test generation process, IPOG-C will call the constraint solver many times, and there is a great chance that some solver calls are identical. Recording the solving history and reusing the results can help reduce the number of solver calls.

4.3 IPOG-C Example

We again use the example SUT used in the AETG example to illustrate how IPOG-C works.

Suppose we are testing a web-based application on different client platforms to see whether it can work normally on different system configurations. Possible factors

---

[1] For a brief introduction to CSP and constraint solving, see Sect. 6.1.

include the operating system (OS), web browser, whether Flash plugin is installed and the proxy configurations. For simplicity, each parameter has two values. The SUT model can be described as follows:

```
OS: Windows, Linux
Browser: IE, Firefox
Flash_Installed: Yes, No
Proxy: None, HTTP
```

Besides, we know the Internet Explorer (IE) browser can only be installed on Windows, so the following constraint is specified:

```
Browser=="IE" -> OS=="Windows"
```

For simplicity, we use p_1, p_2, p_3, p_4 to represent the four parameters, and use 1 and 2 to represent their values. The constraint will be transformed to "$p_2 == 1 \rightarrow p_1 == 1$".

First, we start by randomly reordering the parameter. Suppose the reordered parameters are p_1, p_4, p_2, p_3, which are denoted by $\hat{p}_1$, $\hat{p}_2$, $\hat{p}_3$, $\hat{p}_4$, respectively. So constraint "$p_2 = 1 \rightarrow p_1 = 1$" is changed to "$\hat{p}_3 = 1 \rightarrow \hat{p}_1 = 1$" accordingly. The reason that we reorder the parameters in this way is to help explain how IPOG-C deal with constraints.

Then we initialize the test suite as the set of all valid combinations of the first two parameters. It is easy to see that all combinations of $\hat{p}_1$ and $\hat{p}_2$ are valid, so the initial test suite is:

$$\begin{pmatrix} 1 & 1 \\ 1 & 2 \\ 2 & 1 \\ 2 & 2 \end{pmatrix}.$$

Then, we enter the for-loop in line 3:
Column #3 For $i = 3$, we extend the covering array for $\hat{p}_3$. The set of combinations π is as follows:

$$\pi = \{(1, -, 1, -), (1, -, 2, -), (2, -, 1, -), (2, -, 2, -),$$
$$(-, 1, 1, -), (-, 1, 2, -), (-, 2, 1, -), (-, 2, 2, -)\}.$$

Then, we enter the horizontal extension stage:
Row #1 For the first test case, the number for newly-covered target combinations in π is 2 for both value 1 and 2, so we choose value 1 and append it to the test case, which now becomes $(1, 1, 1, -)$. And then we remove the newly-covered target combinations from π, so it becomes:

$$\pi = \{(1, -, 2, -), (2, -, 1, -), (2, -, 2, -),$$
$$(-, 1, 2, -), (-, 2, 1, -), (-, 2, 2, -)\}.$$

Row #2 For the second test case, the number of newly-covered target combinations in π is 1 for value 1 and 2 for value 2, so we choose value 2 and append it to the test case, which now becomes $(1, 2, 2, -)$. And then, we remove the newly-covered target combinations from π, which now becomes:

$$\pi = \{(2, -, 1, -), (2, -, 2, -),$$
$$(-, 1, 2, -), (-, 2, 1, -)\}.$$

Row #3 For the third test case, if we choose value 1 for $\hat{p}_3$, the partial test case will be $(2, 1, 1, -)$, which violates constraint "$\hat{p}_3 = 1 \rightarrow \hat{p}_1 = 1$", so we can only choose value 2. The resulting test case is $(2, 1, 2, -)$, and π becomes:

$$\pi = \{(2, -, 1, -),$$
$$(-, 2, 1, -)\}.$$

Row #4 For the fourth test case, if we choose value 1 for $\hat{p}_3$, the partial test case will be $(2, 2, 1, -)$, which also violates constraint "$\hat{p}_3 = 1 \rightarrow \hat{p}_1 = 1$", so we can only choose value 2. The resulting test case is $(2, 2, 2, -)$, and π stays unchanged.

Now the test suite is

$$\begin{pmatrix} 1\ 1\ 1 \\ 1\ 2\ 2 \\ 2\ 1\ 2 \\ 2\ 2\ 2 \end{pmatrix},$$

and

$$\pi = \{(2, -, 1, -),$$
$$(-, 2, 1, -)\}.$$

Then we enter the vertical extension stage:

(1) For target combination $(2, -, 1, -)$, we check its validity and find it violates constraint "$\hat{p}_3 = 1 \rightarrow \hat{p}_1 = 1$", so it is marked as unsatisfiable and neglected.
(2) For target combination $(-, 2, 1, -)$, it satisfies the constraint, so we add a new test case $(*, 2, 1, -)$ into the test suite.

After extending the covering array for $\hat{p}_3$, the resulting test suite is as follows:

$$\begin{pmatrix} 1\ 1\ 1 \\ 1\ 2\ 2 \\ 2\ 1\ 2 \\ 2\ 2\ 2 \\ *\ 2\ 1 \end{pmatrix},$$

which is a covering array of strength 2 for $\hat{p}_1, \hat{p}_2, \hat{p}_3$.

Column #4 For $i = 4$, we extend the covering array for $\hat{p}_4$. The set of combinations π is as follows:

$$\pi = \{(1, -, -, 1), (1, -, -, 2), (2, -, -, 1), (2, -, -, 2),$$
$$(-, 1, -, 1), (-, 1, -, 2), (-, 2, -, 1), (-, 2, -, 2),$$
$$(-, -, 1, 1), (-, -, 1, 2), (-, -, 2, 1), (-, -, 2, 2)\}.$$

Then, we enter the horizontal extension stage:

Row #1 For the 1st test case, the number for newly-covered target combinations in π is 2 for both value 1 and 2, so we choose value 1 and append it to the test case, which now becomes $(1, 1, 1, 1)$. And then we remove the newly-covered target combinations from π, so it becomes:

$$\pi = \{(1, -, -, 2), (2, -, -, 1), (2, -, -, 2),$$
$$(-, 1, -, 2), (-, 2, -, 1), (-, 2, -, 2),$$
$$(-, -, 1, 2), (-, -, 2, 1), (-, -, 2, 2)\}.$$

Row #2 For the 2nd test case, the number for newly-covered target combinations in π is 2 for value 1 and 3 for value 2, so we choose value 2 and append it to the test case, which now becomes $(1, 2, 2, 2)$. And then we remove the newly-covered target combinations from π, which now becomes:

$$\pi = \{(2, -, -, 1), (2, -, -, 2),$$
$$(-, 1, -, 2), (-, 2, -, 1),$$
$$(-, -, 1, 2), (-, -, 2, 1)\}.$$

Row #3 For the 3rd test case, the number for newly-covered target combinations in π is 2 for both value 1 and value 2, so we choose value 1 and append it to the test case, which now becomes $(2, 1, 2, 1)$. And then, we remove the newly-covered target combinations from π, which now becomes:

$$\pi = \{(2, -, -, 2),$$
$$(-, 1, -, 2), (-, 2, -, 1),$$
$$(-, -, 1, 2)\}.$$

Row #4 For the 4th test case, the number for newly-covered target combinations in π is 1 for both value 1 and value 2, so we choose value 1 and append it to the test case, which now becomes $(2, 2, 2, 1)$. And then, we remove the newly-covered target combinations from π, which now becomes:

$$\pi = \{(2, -, -, 2),$$
$$(-, 1, -, 2),$$
$$(-, -, 1, 2)\}.$$

Row #5 For the 5th test case, the number for newly-covered target combinations in π is 0 for value 1 and 1 for value 2, so we choose value 2 and append it to the test case, which now becomes $(*, 2, 1, 2)$. And then, we remove the newly-covered target combinations from π, which now becomes:

$$\pi = \{(2, -, -, 2), \\ (-, 1, -, 2)\}.$$

After horizontal growth, the test suite is now:

$$\begin{pmatrix} 1 & 1 & 1 & 1 \\ 1 & 2 & 2 & 2 \\ 2 & 1 & 2 & 1 \\ 2 & 2 & 2 & 1 \\ * & 2 & 1 & 2 \end{pmatrix},$$

and

$$\pi = \{(2, -, -, 2), \\ (-, 1, -, 2)\}.$$

Now we enter the vertical growth stage:

(1) For target combination $(2, -, -, 2)$, it satisfies the constraint. Although test case $(*, 2, 1, 2)$ can be modified to cover this combination, the resulting test case $(2, 2, 1, 2)$ violates the constraint. So we have to add one more test case $(2, *, *, 2)$ to the test suite.

(2) For target combination $(-, 1, -, 2)$, it satisfies the constraint, and the new test case $(2, *, *, 2)$ can be modified to cover the combination. The resulting test case $(2, 1, *, 2)$ also satisfies the constraint, so we accept this modification and no new test case is added.

After this iteration terminates, the covering array becomes:

$$\begin{pmatrix} 1 & 1 & 1 & 1 \\ 1 & 2 & 2 & 2 \\ 2 & 1 & 2 & 1 \\ 2 & 2 & 2 & 1 \\ * & 2 & 1 & 2 \\ 2 & 1 & * & 2 \end{pmatrix}.$$

The test suite is a covering array for all the four parameters.

To finalize, the don't-care values in the covering array needs to be assigned. Since with the presence of constraints, the user may have no idea which value to fill in, and it is possible after some values assigned to parameters with don't-care values, the resulting test case violates the constraints. We again check the validity of each test case with don't-care values, and the solver produces a feasible solution, which could be translated back into a full test case. The resulting test suite is:

$$\begin{pmatrix} 1\ 1\ 1\ 1 \\ 1\ 2\ 2\ 2 \\ 2\ 1\ 2\ 1 \\ 2\ 2\ 2\ 1 \\ 1\ 2\ 1\ 2 \\ 2\ 1\ 2\ 2 \end{pmatrix}.$$

And one more step, we reorder the columns back to the original order of parameters, and the resulting test suite is:

$$\begin{pmatrix} 1\ 1\ 1\ 1 \\ 1\ 2\ 2\ 2 \\ 2\ 2\ 1\ 1 \\ 2\ 2\ 1\ 2 \\ 1\ 1\ 2\ 2 \\ 2\ 2\ 2\ 1 \end{pmatrix}.$$

4.4 IPO Variants

IPO is a very general framework. Since the original algorithm was proposed, there have been a number of extensions to the basic framework.

- *Covering strength.* The original IPO algorithm was used for *pairwise* test generation [3, 7]. It was extended to support t-way test generation by initializing the starting test suite as the set of all the combinations of the first t parameters [4]. Nie et al. [6] and Wang et al. [8, 9] also provided modifications for support of variable-strength covering arrays.
- *Initial parameter order.* The IPOG algorithm [4] suggests putting parameters into an arbitrary order, and some research papers [5, 10] suggest sorting the parameters in nonincreasing order of their domain sizes.
- *Horizontal growth.* In the original IPO algorithm, two horizontal growth algorithms were proposed: (1) By enumerating all possible permutations of parameter values for the new column, and selecting the permutation covering the greatest number of uncovered target combinations. However this strategy is extremely expensive; (2) By assigning each value of parameter p_i once for the first s_i rows, and for the remaining rows, selecting values greedily just as the IPOG algorithm. The special treatment for the first s_i rows is actually unnecessary, and was not used in later IPO-based algorithms. Forbes et al. [2] proposed that if for a certain row, all values for parameter p_i cover no uncovered target combinations, a don't-care value is added

to the row. They also provided an optimization to reduce the computation cost to determine the number of newly-covered target combinations when selecting values for each row.

- *Enumeration of value combinations.* Since the IPOG strategy explicitly enumerates all target combinations, and the number of target combinations may grow exponentially as the covering strength t increases, the cost will be extremely large when k, t and s are large. The IPOG-D algorithm [5] combines the IPOG algorithm with a mathematical construction method called "D-construction", which can double the number of parameters in a three-way covering array [1]. IPOG-D divides the parameters into two groups. It constructs a t-way covering array for the first group and a $(t - 1)$-way covering array for the second group; then it uses D-construction to construct a larger array of all the parameters. If $t > 3$, IPOG-D will use the vertical growth of the IPOG algorithm to cover the uncovered target combinations. By using D-construction, the number of enumerated target combinations can be greatly reduced.

References

1. Chateauneuf, M.A., Colbourn, C.J., Kreher, D.L.: Covering arrays of strength three. Des. Codes Cryptogr. **16**(3), 235–242 (1999)
2. Forbes, M., Lawrence, J., Lei, Y., Kacker, R.N., Kuhn, D.R.: Refining the in-parameter-order strategy for constructing covering arrays. J. Res. Natl. Inst. Stand. Technol. **113**(5), 287–297 (2008)
3. Lei, Y., Tai, K.C.: In-parameter-order: a test generation strategy for pair-wise testing. In: Proceedings of the 3rd IEEE International Symposium on High-Assurance Systems Engineering (HASE'98), IEEE Computer Society, pp. 254–261 (1998)
4. Lei, Y., Kacker, R., Kuhn, D.R., Okun, V., Lawrence, J.: IPOG: A general strategy for t-way software testing. In: Proceedings of the 14th Annual IEEE International Conference and Workshops on the Engineering of Computer-Based Systems (ECBS'07), IEEE, pp. 549–556 (2007)
5. Lei, Y., Kacker, R., Kuhn, D.R., Okun, V., Lawrence, J.: IPOG/IPOG-D: efficient test generation for multi-way combinatorial testing. Softw. Test. Verification Reliab. **18**(3), 125–148 (2008)
6. Nie, C., Xu, B., Shi, L., Dong, G.: Automatic test generation for n-way combinatorial testing. In: Quality of Software Architectures and Software Quality (QoSA-SOQUA), Springer LNCS 3712, pp. 203–211 (2005)
7. Tai, K.C., Lei, Y.: A test generation strategy for pairwise testing. IEEE Trans. Softw. Eng. **28**(1), 109–111 (2002)
8. Wang, Z., Nie, C., Xu, B.: Generating combinatorial test suite for interaction relationship. In: Proceedings of the Fourth International Workshop on Software Quality Assurance (SOQUA 2007), ACM, pp. 55–61 (2007)
9. Wang, Z., Xu, B., Nie, C.: Greedy heuristic algorithms to generate variable strength combinatorial test suite. In: Proceedings of the Eighth International Conference on Quality Software (QSIC'08), IEEE, pp. 155–160 (2008)
10. Yu, L., Lei, Y., Nourozborazjany, M., Kacker, R. N., Kuhn, D. R.: An efficient algorithm for constraint handling in combinatorial test generation. In: Proceedings of the Sixth International Conference on Software Testing, Verification and Validation (ICST'13), IEEE, pp. 242–251 (2013)

Chapter 5
Evolutionary Computation and Metaheuristics

Abstract This chapter describes how to use metaheuristic search and evolutionary algorithms to generate covering arrays for combinatorial testing. They include, among others, genetic algorithms, simulated annealing, tabu search, and particle swarm optimization.

5.1 Methods for Solving Optimization Problems

As we mentioned earlier, it is a difficult problem to find small covering arrays. People often use heuristic search methods to solve it. Such a method can produce a solution within a reasonable amount of time, for many instances of the problem. The AETG algorithm [7] is an example of such methods.

Instead of using a problem-specific heuristic method, we can also use the so-called metaheuristics [20], which provide a general framework for solving a class of optimization problems. Such an optimization problem typically has a number of (discrete or continuous) variables and a cost function. The goal is to find suitable values for the variables such that the cost function gets the minimal or maximal value. Sometimes, we may put some constraints on the variables, which forbid them to take certain values.

A straightforward method for solving (constrained) optimization problems is using simple local search. Such a method starts with a potential solution to the problem; and checks its immediate neighbors (e.g., tables that differ from the current table by one element). Then the method moves to a better solution (often the best among its neighbors). If we are lucky, we can find the best solution in a number of steps. But more often than not, the search process will be trapped in a *local minimum*. In other words, it is stuck in some suboptimal points. At such a point, the value of the cost function is better than that of the neighbors; but it is not the best, if we consider the whole search space. For instance, in Fig. 5.1, g is the global minimum, while p_1 is the local minimum.

To deal with the local minimum problem, we can, for example, repeat the above search process for a number of times, each time with a new random initial array. We may also use some more sophisticated metaheuristics, which come from nature.

© The Author(s) 2014

J. Zhang et al., *Automatic Generation of Combinatorial Test Data*,
SpringerBriefs in Computer Science, DOI 10.1007/978-3-662-43429-1_5

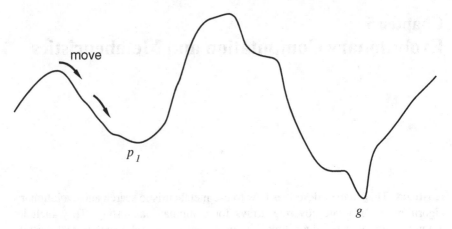

Fig. 5.1 Local minimum

They are sometimes called Nature-Inspired Algorithms [6, 51]. The following are some well-known algorithms for solving difficult optimization problems: Genetic Algorithms, Simulated Annealing, Particle Swarm, Artificial Immune Systems, Ant Colony Optimization, Bees Algorithm, and so on. These algorithms are also called *evolutionary algorithms* [11, 44, 50].

5.2 Applying Evolutionary Algorithms to Test Generation

Quite some of the above-mentioned evolutionary algorithms (EAs) have been applied to the generation of covering arrays (CAs). See for example, [4, 9, 36, 45].

We may distinguish between two different ways of applying EAs to the generation of CAs. One approach is *local*, i.e., embed an EA into some greedy method like AETG. Specifically, we use an EA to find the new test case, in Algorithm 1 (Sect. 3.1). See Fig. 5.2a. We try to determine the last row, using some EA.

The other approach is *global*, in the sense that, we apply an EA to find the whole covering array. See Fig. 5.2b. In this case, we usually assume that the array has some fixed number of rows, e.g., N.

Fig. 5.2 Different ways of applying evolutionary algorithms

(a)
1 1 1 1
1 2 2 2
2 2 1 1
? ? ? ?

(b)
? ? ? ?
? ? ? ?
? ? ? ?
? ? ? ?
? ? ? ?
? ? ? ?

5.3 Genetic Algorithms

The genetic algorithm (GA) is one of the oldest and most popular evolutionary algorithms. GAs have been used in a wide variety of applications. There are many papers and books about genetic algorithms. See for example, [10, 26, 28, 35, 40]. The basic idea is to simulate the process of natural selection.

Initially, we have a set of randomly generated candidates. The main part of a GA is an iterative process. The population in each iteration is called a generation. It can evolve through *inheritance, mutation, selection* and *crossover*.

When using GA to solve a practical optimization problem, we need to represent each candidate solution as a *chromosome*, which is a sequence/string of values. We also need a *fitness function*, telling us how good a candidate solution is. The goal is to *maximize* the value of the fitness function.

Genetic algorithms have been used by various researchers to find covering arrays and other combinatorial designs.

5.3.1 Applying GA to Obtain the Whole Covering Array

An early work in this area is outlined in Ghazi and Ahmed [21]. In the approach, each chromosome consists of a number of test cases (called *configurations* in Ghazi and Ahmed [21]). The fitness function for evaluating a chromosome is defined to be the number of distinct pairwise interaction configurations covered by all of the chromosome's configurations, divided by the total number of possible pairwise interaction configurations $|\Phi_2|$. Here, Φ_2 is the set of all possible pairwise interaction configurations.

For example, suppose the SUT has 3 parameters, each of which can only take the value 1 or 2. Then Φ_2 is:

$$\{\{1, 1, -\}, \{1, 2, -\}, \{2, 1, -\}, \{2, 2, -\}, \{1, -, 1\}, \{1, -, 2\},$$

$$\{2, -, 1\}, \{2, -, 2\}, \{-, 1, 1\}, \{-, 1, 2\}, \{-, 2, 1\}, \{-, 2, 2\}\}$$

And its size is 12.

Given a chromosome C_1 which has two configurations: $\{\{1, 2, 2\}, \{1, 1, 2\}\}$. The set of pairwise interaction configurations covered by C_1 can be calculated as follows. For $i = 1, 2$, let N_i denote the set of distinct pairwise interaction elements covered by configuration i. Then

$$N_1 = \{\{1, 2, -\}, \{1, -, 2\}, \{-, 2, 2\}\}$$
$$N_2 = \{\{1, 1, -\}, \{1, -, 2\}, \{-, 1, 2\}\}$$

The total number of distinct pairwise configurations covered by the chromosome C_1 is 5 (because there is a common element in N_1 and N_2). So the fitness function value of C_1 is $5/12$.

Recently, McCaffrey [32, 34] implemented (in C#) a test data generation method, which is also based on genetic algorithms. The technique/tool is called GAPTS. It first maps all possible parameter values to consecutive integers, e.g., $0, 1, 2, \ldots, 6$. They serve as the gene values. A chromosome is represented as an array of integers. For instance, suppose the SUT has three parameters. Then a test case is represented by a sequence of three integers, and the array $[0, 3, 6, 1, 2, 5, 1, 4, 5, 0, 2, 6]$ is a chromosome, which denotes a test suite of four test cases. The fitness function is simply defined as the total number of distinct pairs in the chromosome representation of test cases.

McCaffrey discussed some issues in the implementation, such as the population size and mutation rate. The population size is set to 20, and the mutation rate is fixed as 0.001. His experimentation indicates that, mutation rates larger than 0.001 tend to modify good solutions too often, which result in slow convergence; mutation rates smaller than 0.001 often lead to premature convergence to nonoptimal solutions.

McCaffrey compared the tool GAPTS with other tools like AETG and PICT. In general, GAPTS produces comparable or smaller test suites, but it requires much longer time.

The above works only deal with pairwise testing. And they apply the GA to the whole array. Alternatively, it is also possible to apply GA to obtain a single test case.

5.3.2 Applying GA to Obtain a Good Test Case

Recall that AETG is a greedy algorithm which constructs a test suite by repeatedly adding a test case that covers as many uncovered interactions as possible. In each iteration, it creates many candidate test cases and selects the one that covers the largest number of newly covered combinations. Clearly, this can be regarded as an optimization problem.

Shiba et al. [43] extended the basic AETG algorithm with the genetic algorithm, and obtained the test generation procedure AETG-GA.

In AETG-GA, each test case is treated as a chromosome. The fitness function $f(s)$ for a test case s is the number of new t-way combinations that are not covered by the given test suite but are covered by s.

One feature of AETG-GA is that, in each generation, the best σ chromosomes in the population are kept and they survive to the next generation. This is called the elite strategy in Shiba et al. [43].

5.4 Simulated Annealing

Simulated Annealing (SA) [31] is a probabilistic method for solving global optimization problems. It is inspired from the annealing technique in metallurgy, which involves heating and controlled cooling of materials. The annealing process results in low-energy structure of crystals. We can simulate this process and obtain an algorithm to *minimize* some cost or energy function. The cost corresponds to the energy of some crystalline solid.

The SA algorithm takes random walks in the search space, looking for points with low energies. We start with an initial candidate solution s_0, and repeatedly move to new points. At each step, we randomly generate an alternative candidate solution and measure its cost/energy. If the new energy E' is lower than the current one E, we move to that point and update the candidate solution; otherwise, we move to the new point, with probability $e^{-(E'-E)/(kT)}$. Here, T is the "temperature", and k is Boltzmann's constant. Initially the temperature is high, making the candidate solution more likely to change; then the temperature is lowered very slightly according to some cooling schedule.

Some implementations of SA are available. For example, an implementation is contained in the GNU Scientific Library (GSL),[1] which is a numerical library for C and C++ programmers.

Stevens [46] uses SA to construct small covering arrays (transversal covers). The cost function is the number of uncovered pairs. The algorithm starts with a random matrix. Each move is just selecting a random row and changing a random value in that row. The proposed move is accepted with a probability that decreases exponentially with time. The matrix cools until the probability becomes zero. If we obtain a matrix whose cost is zero, we have found a covering array.

Typically, before we apply the SA algorithm, we have to guess an initial size of the covering array. During the annealing process, the number of rows is fixed (as some constant N). If the algorithm finds a covering array, the size is decreased by one and a new run of the SA process is started.

Garvin et al. [17] studied how to improve the SA algorithm to deal with constraints. In addition, they discussed how to determine the size of the covering array. That is called the *outer search*, which is concerned with the minimization of N. It takes an upper bound and a lower bound on the size of the covering array, and performs a binary search within this range. The *inner search* performs SA, and explores the space of Nk arrays by changing one entry at a time, guided by the cost function.

Recently, Torres-Jimenez and Rodriguez-Tello [47] described an improved implementation of the SA algorithm, called ISA, for constructing *binary* covering arrays of strengths 3–6. Using ISA, they obtained 104 new bounds.

ISA has two key features:

---

[1] http://www.gnu.org/software/gsl/.

- a heuristic to generate good initial solutions. Typically, the initial array is randomly generated, and then the SA process is started. But ISA uses an initialization procedure that keeps a balanced number of symbols in each column of the array.
- a compound neighborhood function, which combines two carefully designed neighborhood relations. The neighborhood function specifies the set of potential solutions that can be reached (in one step) from the current solution. For every potential solution/array A, it assigns a set of neighboring solutions, which is called the neighborhood of A.

Simulated annealing can be combined with other techniques. Cohen et al. [8] proposed a method for constructing covering arrays, which combines SA with the algebraic approach (based on mathematical results). Using this method, the authors obtained new bounds on the size of some strength three covering arrays.

Rodriguez-Cristerna and Torres-Jimenez [41] proposed a hybrid approach to the construction of mixed covering arrays, which combines SA with a Variable Neighborhood Search function. It is called SA-VNS.

5.5 Particle Swarm Optimization

Particle Swarm Optimization (PSO) [30, 42] is a method for optimizing continuous nonlinear functions. For example, finding an assignment to variables x and y (within a certain range), so as to minimize a function like $x^2 + xy - 3$. The algorithm simulates some social behaviors of humans and animals (e.g., that of flying birds).

In a PSO algorithm, there are also a population of candidate solutions (called a swarm of particles). The position of each particle represents a possible solution. However, different from other evolutionary methods like genetic algorithms, a PSO has a velocity vector. Each particle moves in the search space with some velocity.

PSO is an iterative process. In each iteration, each particle's current velocity is first updated; then its position is updated using the new velocity.

There are several parameters in the PSO algorithm, e.g., the number of particles, the neighborhood size (that is, how many nearest neighbors can influence an article), the maximum velocity. The values of these parameters have quite some influence on the performance of the algorithm. Various researchers have studied how to select good parameter values for PSO. See for example, [16, 38, 48].

There are some open-source implementations of PSO. For example, SwarmOps (for numerical and heuristic optimization) is available at http://www.hvass-labs.org/projects/swarmops/

Ahmed et al. [1, 2] applied PSO to the construction of covering arrays. Their tool Particle Swarm-based t-way Test Generator (PSTG) can be used to generate uniform and variable strength covering arrays. It compares favorably with other tools like PICT, Jenny, IPOG/IPOG-D etc., in terms of the size of the generated test suite.

5.6 Tabu Search

Tabu search [22–25] is an effective metaheuristic search method for solving optimization problems. Tabu (taboo) means "forbidden", "prohibited", "banned". Tabu search improves the performance of local search methods by remembering the recently visited solutions. If a certain region of the search space has been visited recently, then it is marked as tabu, and the search process is not allowed to visit the region again (in the near future).

Nurmela [37] uses tabu search to find small covering arrays of strength t. As usual, the problem is regarded as an optimization problem, whose cost function is the number of uncovered t-combinations.

The approach starts with a random matrix and moves to neighbors repeatedly until a matrix with zero cost (covering array) is found or the number of moves has reached a prescribed limit. At each step, the approach selects one uncovered t-combination at random. It checks which rows require only the change of one element such that the row covers the selected combination. These changes are the moves to be considered. The cost change corresponding to each move is calculated and the move leading to the smallest cost is selected, provided that the move is not tabu. If there are several equally good non-tabu moves, a move is selected randomly. The tabu condition prevents changing an element of the matrix, if it has been changed during the last T moves. Here T is some small positive integer (e.g., $1 \leq T \leq 10$).

Recently, Walker and Colbourn [49] introduced an effective tabu search method for finding certain covering arrays. Such covering arrays have a compact representation which employs "permutation vectors". With the new technique, improved covering arrays of strength 3, 4, and 5 have been found.

Gonzalez-Hernandez and Torres-Jimenez [27] proposed another tabu search method (called MiTS) for constructing mixed covering arrays. This approach uses a mixture of neighborhood functions and a fine tuning process to improve the performance of tabu search.

Let N denote the number of rows (test cases), and k denote the number of columns (parameters). Let s denote a candidate solution, and (i, j) is an arbitrary position in the matrix (the ith row, the jth column). Suppose the jth parameter can take v_j different values.

Instead of using a single neighborhood function $F(s)$, MiTS uses the following three functions:

- $F_1(s)$: A neighbor is obtained by changing the symbol/value at position (i, j). Using this function, s has $v_j - 1$ possible neighbors.
- $F_2(s)$: A neighbor is obtained by changing the symbol/value at any position in column j – but only one position is allowed to change each time. Using this function, s has $(v_j - 1) * N$ possible neighbors.
- $F_3(s)$: The change can be made to any single element in the matrix. Using this function, s has $(v_j - 1) * N * k$ possible neighbors.

5.7 Other Methods

During the past 20 years, many evolutionary algorithms have been developed. In this section, we briefly describe several such algorithms that have been used for finding combinatorial test suites.

Ant Colony optimization (ACO) [13–15] is an algorithm inspired by the behavior of ants. Each ant is very small and its ability is very limited. However, by working together, the ants are very good at performing various tasks. For example, they can often find a short path to a food source. Thus a natural application of ACO is to find optimal paths, e.g., solving the traveling salesman problem (TSP) [12].

In addition to AETG-GA, Shiba et al. [43] also proposed a test generation procedure called AETG-ACA, which extends the basic AETG algorithm with the ant colony algorithm.

Chen et al. [5] also extended the one-test-at-a-time strategy to build test suites. When generating a single test case, they adopted the ant colony system (ACS) strategy, which is an effective variant in the ACO family.

The Bees algorithm, proposed by Duc Truong Pham et al. [39], is a population-based search algorithm for solving optimization problems. It is inspired by the food foraging behaviour of swarms of honey bees.

Independently, Dervis Karaboga and his collaborators proposed the Artificial Bee Colony algorithm (http://mf.erciyes.edu.tr/abc/index.htm).

McCaffrey [33] implemented the Bee Colony algorithm to find pair-wise test suite. His tool SBC compares favorably with other tools like PICT and AETG on the size of the test suite. But its running time is much longer. For some instance, it needs around 20 min to find a solution, while PICT finishes within a few seconds.

Harmony search [18, 19] is another evolutionary algorithm for solving optimization problems, which is inspired by musical processes. Alsewari and Zamli [3] proposed a test generation method based on harmony search algorithm, called Harmony Search Strategy (HSS). HSS addresses the support for high interaction strength and the support for constraints. HSS was implemented in Java. Compared with other tools like PICT, Density and IPOG, HSS is shown to be able to obtain smaller test suites.

References

1. Ahmed, B.S., Zamli, K.Z.: A variable strength interaction test suites generation strategy using particle swarm optimization. J. Syst. Softw. **84**(12), 2171–2185 (2011)
2. Ahmed, B.S., Zamli, K.Z., Lim, C.P.: Application of particle swarm optimization to uniform and variable strength covering array construction. Appl. Soft Comput. **12**(4), 1330–1347 (2012)
3. Alsewari, A.R.A., Zamli, K.Z.: Design and implementation of a harmony-search-based variable-strength t-way testing strategy with constraints support. Inf. Softw. Technol. **54**(6), 553–568 (2012)

4. Bryce, R.C., Colbourn, C.J.: One-test-at-a-time heuristic search for interaction test suites. In: Proceedings of 9th Conference on Genetic and Evolutionary Computation (GECCO'07), pp. 1082–1089 (2007)
5. Chen, X., Gu, Q., Li, A., Chen, D.: Variable strength interaction testing with an ant colony system approach. In: Proceedings of the 16th Asia-Pacific Software Engineering Conference (APSEC), pp. 160–167 (2009)
6. Chiong, R. (ed.): Nature-Inspired Algorithms for Optimisation. Springer, Berlin (2009)
7. Cohen, D.M., Dalal, S.R., Fredman, M.L., Patton, G.C.: The aETG system: an approach to testing based on combinatorial design. IEEE Trans. Softw. Eng. **23**(7), 437–444 (1997)
8. Cohen, M.B., Colbourn, C.J., Ling, A.C.H.,: Augmenting simulated annealing to build inter-action test suites. In: Proceedings of the 14th International Symposium on Software Reliability Engineering (ISSRE 2003), pp. 394–405 (2003)
9. Cohen, M.B., Gibbons, P.B., Mugridge, W.B., Colbourn, C.J.: Constructing test suites for inter-action testing. In: Proceedings of the 25th International Conference on Software Engineering (ICSE), pp. 38–48 (2003)
10. Davis, L. (ed.): Handbook of Genetic Algorithms. Van Nostrand Reinhold, New York (1991)
11. De Jong, K.: Evolutionary Computation. The MIT Press, Cambridge (2002)
12. Dorigo, M., Maniezzo, V., Colorni, A.: Ant system: optimization by a colony of cooperating agents. IEEE Trans. Syst. Man Cybern. Part B **26**(1), 29–41 (1996)
13. Dorigo, M., Stützle, T.: Ant Colony Optimization. The MIT Press, Cambridge (2004)
14. Dorigo, M., Blum, C.: Ant colony optimization theory: a survey. Theor. Comput. Sci. **344**(2–3), 243–278 (2005)
15. Dorigo, M., Birattari, M., Stützle, T.: Ant colony optimization. IEEE Comput. Intell. Mag. **1**(4), 28–39 (2006)
16. Eberhart, R.C., Shi, Y.: Comparing inertia weights and constriction factors in particle swarm optimization. Proc. Congr. Evol. Comput. **1**, 84–88 (2000)
17. Garvin, B., Cohen, M., Dwyer, M.: Evaluating improvements to a meta-heuristic search for constrained interaction testing. Empir. Softw. Eng. **16**(1), 61–102 (2011)
18. Geem, Z.W.: Recent Advances in Harmony Search Algorithm. Springer, Berlin (2010)
19. Geem, Z.W., Kim, J.-H., Loganathan, G.V.: A new heuristic optimization algorithm: harmony search. Simulation **76**(2), 60–68 (2001)
20. Gendreau, M., Potvin, J.-Y. (eds.): Handbook of Metaheuristics, 2nd edn. Springer, New York (2010)
21. Ghazi, S.A., Ahmed, M.A.: Pair-wise test coverage using genetic algorithms. Proc. Congr. Evol. Comput. (CEC) **2**, 1420–1424 (2003)
22. Glover, F.: Future paths for integer programming and links to artificial intelligence. Comput. Oper. Res. **13**(5), 533–549 (1986)
23. Glover, F.: Tabu search—part 1. ORSA J. Comput. **1**(2), 190–206 (1989)
24. Glover, F.: Tabu search—part 2. ORSA J. Comput. **2**(1), 4–32 (1990)
25. Glover, F., Taillard, E., de Werra, D.: A user's guide to tabu search. Ann. Oper. Res. **41**(1), 1–28 (1993)
26. Goldberg, D.E.: Genetic Algorithms in Search, Optimization, and Machine Learning. Addison-Wesley Longman Publishing Co., Inc., Boston (1989)
27. Gonzalez-Hernandez, L., Torres-Jimenez, J.: MiTS: a new approach of tabu search for con-structing mixed covering arrays. In: Proceedings of the 9th Mexican International Conference on Artificial Intelligence (MICAI), LNCS 6438, pp. 382–393 (2010)
28. Haupt, R.L., Haupt, S.E.: Practical Genetic Algorithms. Wiley, New York (2004)
29. Karaboga, D.: Artificial bee colony algorithm. Scholarpedia **5**(3), 6915 (2010). http://www.scholarpedia.org/article/Artificial_bee_colony_algorithm
30. Kennedy, J., Eberhart, R.: Particle swarm optimization. Proc. IEEE Int. Conf. Neural Netw. **4**, 1942–1948 (1995)
31. Kirkpatrick, S., Gelatt Jr, C.D., Vecchi, M.P.: Optimization by simulated annealing. Science **220**(4598), 671–680 (1983)

32. McCaffrey, J.D.: Generation of pairwise test sets using a genetic algorithm. In: Proceedings of the 33rd Annual IEEE International Computer Software and Applications Conference (COMP-SAC), Vol. 1, pp. 626–631 (2009)
33. McCaffrey, J.D.: Generation of pairwise test sets using a simulated bee colony algorithm. In: Proceedings of the IEEE International Conference on Information Reuse and Integration (IRI) pp. 115–119, IEEE (2009)
34. McCaffrey, J.D.: An empirical study of pairwise test set generation using a genetic algorithm. In: Proceedings of the 7th International Conference on Information Technology: New Generations (ITNG), pp. 992–997, (2010)
35. Mitchell, M.: An Introduction to Genetic Algorithms. The MIT Press, Cambridge (1998)
36. Nie, C., Wu, H., Liang, Y., Leung, H., Kuo, F.-C., Li, Z.: Search based combinatorial testing. In: Proceedings of the Asia-Pacific Software Engineering Conference (APSEC), pp. 778–783 (2012)
37. Nurmela, K.J.: Upper bounds for covering arrays by tabu search. Discret. Appl. Math. **138**(1–2), 143–152 (2004)
38. Pedersen, M.E.H.: Good parameters for particle swarm optimization. Hvass Laboratories Technical Report HL1001 (2010). http://www.hvass-labs.org
39. Pham, D.T., Ghanbarzadeh, A., Koc, E., Otri, S., Rahim, S., Zaidi, M.: The Bees algorithm—a novel tool for complex optimisation problems. In: Proceedings of IPROMS Conference, pp. 454–461 (2006)
40. Reeves, C.R., Rowe, J.E.: Genetic Algorithms: Principles and Perspectives: A Guide to GA Theory. Kluwer Academic, Dordrecht (2002)
41. Rodriguez-Cristerna, A., Torres-Jimenez, J.: A simulated annealing with variable neighborhood search approach to construct mixed covering arrays. Electron. Notes Discret. Math. **39**(1), 249–256 (2012)
42. Shi, Y., Eberhart, R.C.: A modified particle swarm optimizer. In: Proceedings of IEEE International Conference on Evolutionary Computation, pp. 69–73 (1998)
43. Shiba, T., Tsuchiya, T., Kikuno, T.: Using artificial life techniques to generate test cases for combinatorial testing. In: Proceedings of the 28th Annual International Computer Software and Applications Conference (COMPSAC), vol. 1, pp. 72–77 (2004)
44. Simon, D.: Evolutionary Optimization Algorithms. Wiley, Hoboken (2013)
45. Stardom, J.: Metaheuristics and the search for covering and packing arrays. Master's thesis, Simon Fraser University, Canada (2001)
46. Stevens, B.: Transversal Covers and Packings, Ph.D. thesis, University of Toronto, Toronto, Canada (1998)
47. Torres-Jimenez, J., Rodriguez-Tello, E.: New bounds for binary covering arrays using simulated annealing. Inf. Sci. **185**(1), 137–152 (2012)
48. Trelea, I.C.: The particle swarm optimization algorithm: convergence analysis and parameter selection. Inf. Process. Lett. **85**(6), 317–325 (2003)
49. Walker II, R.A., Colbourn, C.J.: Tabu search for covering arrays using permutation vectors. J. Stat. Planning Inf. **139**(1), 69–80 (2009)
50. Whitley, D.: An overview of evolutionary algorithms: practical issues and common pitfalls. Inf. Softw. Technol. **43**(14), 817–831 (2001)
51. Zang, H., Zhang, S., Hapeshi, K.: A review of nature-inspired algorithms. J. Bionic Eng. **7**, S232–S237 (2010) (Suppl, Elsevier)

Chapter 6
Backtracking Search

Abstract This chapter describes how to use backtracking search to find covering arrays and orthogonal arrays. The basic algorithms are presented, as well as various heuristics which improve the efficiency of the search. Different from the approaches described in the previous chapters, backtracking search is able to find the smallest test suite.

6.1 Constraint Satisfaction Problems

Constraint satisfaction, in its basic form, involves finding a value for each one of a set of variables so that all of the given constraints are satisfied. The formal definition of a Constraint Satisfaction Problem (CSP) is as follows:

Definition 6.1 A constraint satisfaction problem $\mathscr{P}$ is a triple $\mathscr{P} = \langle X, D, C \rangle$, where X is an n-tuple of variables $X = \langle x_1, x_2, \ldots, x_n \rangle$, D is an n-tuple of domains $D = \langle D_1, D_2, \ldots, D_n \rangle$, C is a collection of constraints $C = \langle C_1, C_2, \ldots, C_m \rangle$. Each variable x_i may take a value from the corresponding D_i. Each constraint C_j involves some subset of the variables and specifies the allowable combinations of values for that subset.

To solve a CSP, we need to find a value $v_i \in D_i$ for each variable x_i, such that all the constraints in C hold.

CSP is a general framework for many specific problems. For example, the graph (vertex) coloring problem can be regarded as a CSP. Here the variables are the colors of the vertices; and for each pair of adjacent vertices, we have a constraint of the form $x_i \neq x_j$. Here x_i and x_j correspond to the colors of the two vertices.

For more information about CSPs, the reader may refer to the handbook [7].

An important special case of CSP is the Boolean/propositional satisfiability problem (SAT): Given a Boolean formula $F(x_1, x_2, \ldots, x_n)$, determine if it is satisfiable. If yes, return the values of variables which make F true.

A Boolean formula is commonly formulated in *Conjunctive Normal Form* (CNF). Now we briefly describe the meaning of CNF. A *literal* is a Boolean variable or its negation. For instance, x, $\neg y$ are two literals. A *clause* is the disjunction (logical

© The Author(s) 2014
J. Zhang et al., *Automatic Generation of Combinatorial Test Data*,
SpringerBriefs in Computer Science, DOI 10.1007/978-3-662-43429-1_6

OR) of one or more literals. The following are two clauses: $x \vee y$, $x \vee \neg y \vee z$. A boolean expression/formula is said to be in Conjunctive Normal Form (CNF), if it is the conjunction (logical AND) of clauses. Here is an example:

$$(x \vee \neg y) \wedge (\neg x \vee y)$$

SAT is the first known NP-complete problem. Despite its theoretical hardness, efficient algorithms have been developed for solving SAT.

6.1.1 Backtracking Algorithms for Solving CSPs

Backtracking is the fundamental complete search methodology for solving CSPs of finite domains. It is a refinement of the brute force approach, *systematically* searching for a solution among all possibilities.

The basic procedure of backtracking is to gradually extend a partial solution by choosing values for variables until all constraints are satisfied, in which case a solution is found, or until all possibilities are exhausted. If the current partial solution cannot be consistently extended, backtracking is needed–the last choice is cancelled and a new choice is tried.

During the search, a partial solution looks like this: $\langle x_1 = 4, x_3 = 2 \rangle$. To extend the partial solution, we select a variable which does not have a value yet, say x_6. Then we choose a value for it, e.g., let $x_6 = 8$. If this assignment does not result in any inconsistency, we take it, get a new partial solution which has 3 assignments; and then we try to extend it further. On the other hand, if no good value can be assigned to x_6, we need to backtrack, and make a different choice (assigning a different value to x_3). Then we get a different partial solution, e.g., $\langle x_1 = 4, x_3 = 6 \rangle$.

Unlike brute force, which simply enumerates and tests all candidate solutions, backtracking checks if any constraint is violated each time a variable is assigned, so that some candidate solutions can be abandoned at an earlier stage. The backtracking search process is often represented as a search tree, where each branch corresponds to a choice (i.e., variable assignment), and each path represents a candidate partial solution.

6.1.2 Constraint Propagation

When solving a CSP via backtracking search, exploring the whole space of variable instantiations is too expensive. Constraint propagation is an important technique to reduce the number of candidate values for variables by detecting local inconsistency. As a result, some branches with no solution in the search tree can be pruned earlier, making the problem easier to solve.

Fig. 6.1 Search for an idempotent Latin square

(a)					**(b)**					**(c)**			
0	?	?	?		0	2	?	?		0	2	3	1
?	1	?	?		?	1	?	?		3	1	0	2
?	?	2	?		?	?	2	?		1	3	2	0
?	?	?	3	$\Rightarrow$	?	?	?	3	$\Rightarrow$	2	0	1	3

The constraint propagation algorithm proceeds as follows. Constraints are used actively to reduce domain sizes by filtering values that cannot take part in any solution. Once the domain of a variable is reduced, more consistencies might be triggered. Therefore the algorithm recomputes the candidate value sets of all its dependent variables by detecting inconsistency w.r.t. the constraints. This process continues recursively until there is no more propagation. It is worth noticing that constraint propagation preserves all solutions to the original problem. There are several methods for constraint propagation, with different power and cost.

Example 6.1 Let us see how we can find an idempotent Latin square of order 4. Assume the cell at row i and column j is $L(i, j)$ ($0 \leq i, j < 4$). Then an *idempotent* Latin square has the property that, for each i, $L(i, i) = i$. So, initially, the matrix is like that of Fig. 6.1a.

Now let us choose a value for the second cell on the first row, i.e., $L(0, 1)$. According to the properties of Latin squares, this cell can not take the value 0 or 1; it can only be 2 or 3. Assume that $L(0, 1) = 2$, as shown by Fig. 6.1b. After this choice, the domains of other variables become smaller. For instance, $L(0, 2)$ can only be 1 or 3.

If the reasoning of the algorithm is powerful, we may find that there is only one value for $L(0, 2)$. It can not be 1. Otherwise, we have $L(0, 3) = 3$. This will contradict with the fact that $L(3, 3) = 3$. So we conclude that $L(0, 2) = 3$. After similar reasoning (constraint propagation), we get the solution as shown by Fig. 6.1c.

6.1.3 Symmetry Breaking

For a constraint solving problem, a **symmetry** is a one to one mapping (bijection) on variables that preserves solutions and non-solutions. We say two solutions or search states are **isomorphic** if there is a symmetry mapping one of them to the other. Since isomorphism is caused by symmetry, in the remainder of the chapter, we use symmetry breaking and isomorphism elimination without distinction.

It has long been recognized that symmetries can cause significant problems to systematic solvers that unnecessarily explore redundant parts of the search tree. As a result, symmetry breaking becomes a vital technique in CSP. The goal of symmetry breaking is to avoid exploring as much as possible two isomorphic search states, since the results in both cases must be the same.

There are mainly two approaches to symmetry breaking for a CSP problem. One is to statically add symmetry breaking constraints before search starts, thereby excluding some isomorphic solutions while leaving at least one solution in each symmetric equivalence class. The other is to break symmetry dynamically during search, adapting the search procedure appropriately. The static approach is easy to understand and implement. In particular, it is very cost-effective for matrix models. The symmetry breaking techniques discussed in this chapter are basically classified as the static approach.

6.1.4 Constraint Solving Tools

Many tools have been developed for solving constraints. They are called constraint solvers or constraint programming systems. For instance,

- GeCode (http://www.gecode.org/) is a toolkit for developing constraint-based systems and applications. It can also be used as a library.
- Minion (http://minion.sourceforge.net/) is a fast scalable constraint solver.
- Choco (http://www.emn.fr/z-info/choco-solver/) is a library for constraint solving and constraint programming, implemented in Java.

All of them are open source.

There are many efficient SAT solvers available today, e.g., MiniSat (http://www.minisat.se/). For more information about SAT solving and related tools, see for example, http://www.satlive.org/ and http://www.satisfiability.org/.

6.2 Constraint Solving and the Generation of Covering Arrays

In the previous chapters, we have seen that, to deal with constraints in combinatorial testing, we need to perform some kind of constraint solving or SAT solving. Such techniques have been integrated into the algorithms IPOG-C and AETG-SAT. When implementing AETG-SAT, Cohen et al. [1] used the tool MiniSat. The ACTS tool, which implements IPOG-C [11], uses the Choco library.

In addition, we may transform the whole problem of generating covering arrays into a constraint satisfaction problem (CSP) or SAT. Hnich et al. [3] developed several constraint programming models of the problem. They exploit such techniques as global constraints and symmetry-breaking constraints. In particular, they use compound variables to represent tuples of variables, which allows better propagation. Improved bounds were obtained using this approach, with the help of ILOG Solver 6.0.

Hnich et al. [3] also experimented with the SAT-encoding of the CA generation problem. They used a SAT solver that is based on local search. They noted that this approach is able to find improved bounds for larger problem instances, and the best model for local search is not necessarily the best model for backtrack search.

An optimal covering array $CA(24, 2^{10}, 4)$ is given in [3].

A third way of using constraint solving technology is to design special algorithms. In the following, we shall elaborate on backtracking search algorithms for finding covering arrays and orthogonal arrays.

6.3 Backtracking Search for Generating Covering Arrays

The Covering Array (CA) generation problem can be naturally modeled as a CSP. Without loss of generality, assume we are trying to find a $CA(N, d_1 \cdot d_2 \cdots d_k, t)$ where $d_1 \geq d_2 \geq \cdots \geq d_k$. A CA can be displayed as a 2-dimensional $N \times k$ matrix, where each entry in the matrix is called a *cell*. We use $ce_{r,c}$ to denote the cell at row r and column c. The set $\{ce_{r,c} | 1 \leq r \leq N, 1 \leq c \leq k\}$ is the set of problem variables in the CSP, and each variable $ce_{r,c}$ has domain $\{0, \ldots, d_c - 1\}$. The restrictions on value combinations serve as problem constraints.

6.3.1 Preprocessing

In the first t columns of the matrix, there are $b = d_1 d_2 \ldots d_t$ possible combinations of values. We can make the first b rows contain each value combination once by swapping the rows of the matrix without changing the test suite. Thus we can initialize this $b \times t$ sub-array (which is called an **init-block**) in advance. For example, in Fig. 6.2, the init-block of an instance of $CA(6, 2^5, 2)$ is the framed sub-array of size 4×2.

6.3.2 The Search Procedure

The framework of the backtracking algorithm is described as a recursive procedure in Algorithm 4. For a given covering array number N, we try to assign each cell of the $N \times k$ matrix until all requirements of covering arrays are satisfied. The parameter *pSol* and *Fmla* represent the current partial solution (assignments to cells) and the constraints respectively. Each time a variable is assigned, the constraint propagation function **BPropagate** inspects the constraints to check if any constraint

Fig. 6.2 Init-block of $CA(6, 2^5, 2)$

$$
\begin{array}{|cc|cccc}
\hline
0 & 0 & 0 & 0 & 0 \\
0 & 1 & 0 & 0 & 0 \\
1 & 0 & 0 & 1 & 1 \\
1 & 1 & 1 & 0 & 1 \\
\hline
0 & 0 & 1 & 1 & 1 \\
1 & 1 & 1 & 1 & 0 \\
\end{array}
$$

implies another value (returns TRUE) or if there is contradiction (returns FALSE). The function Chk_Cons checks whether all the t-tuples have been covered in the columns $\{1, \ldots, c\}$ if all the cells in column c have been assigned values. The function CELL_Selection chooses an unassigned cell and function Val_Set_Gen generates the candidate value set S_x of a cell.

Algorithm 4 Backtracking Search BSrh($pSol$, $Fmla$)

1: **if** !BPropagate($pSol$, $Fmla$) **then**
2:    return **FALSE**;
3: **end if**
4: **if** !Chk_Cons($pSol$, $Fmla$) **then**
5:    return **FALSE**;
6: **end if**
7: **if** every cell is assigned **then**
8:    return **TRUE**;
9: **end if**
10: x = CELL_Selection($pSol$);
11: S_x = Val_Set_Gen(x, $pSol$, $Fmla$);
12: **for** each value u in set S_x **do**
13:    **if** BSrh($pSol \cup \{x = u\}$, $Fmla$) **then**
14:       return **TRUE**;
15:    **end if**
16: **end for**
17: return **FALSE**;

6.3.3 Exploiting Symmetries in CA

Obviously two CAs with the same parameters are isomorphic if one can be obtained from the other by a finite sequence of row permutations, column permutations and permutations of symbols in each column.

Example 6.2 Figure 6.3 illustrates three instances of CA($6, 2^5, 2$). Matrix B is obtained from Matrix A by swapping the last two columns, and Matrix C is obtained from Matrix A by swapping the first row with the fifth row. They are isomorphic to each other.

Fig. 6.3 Three isomorphic instances of CA($6, 2^5, 2$)

(a)
```
0 0 0 0 0
0 1 0 0 0
1 0 0 1 1
1 1 1 0 1
0 0 1 1 1
1 1 1 1 0
```

(b)
```
0 0 0 0 0
0 1 0 0 0
1 0 0 1 1
1 1 1 1 0
0 0 1 1 1
1 1 1 0 1
```

(c)
```
0 0 1 1 1
0 1 0 0 0
1 0 0 1 1
1 1 1 0 1
0 0 0 0 0
1 1 1 1 0
```

6.3.3.1 Row and Column Symmetries

Flener et al. [2] identified an important class of symmetries in constraint programming, arising from matrices of decision variables where rows and columns can be swapped. In general, an $m \times n$ matrix has $m! \times n! - 1$ symmetries caused by permutations of rows and columns excluding the identity. It is not practical to generate a super-exponential number of constraints so as to eliminate all the symmetries. Flener et al. found that imposing lexicographic order on both the rows and the columns can break symmetries efficiently, while bringing in only a linear number of constraints.

Lexicographic order is recursively defined as follows:

Definition 6.2 (*Lexicographic Order*) Two vectors $X = [x_1, x_2, \ldots, x_n]$ and $Y = [y_1, y_2, \ldots, y_n]$ have a lexicographic order $X \leq_{\text{lex}} Y$ if

1. If $n = 1$, $x_1 \leq y_1$;
2. If $n > 1$, $x_1 < y_1 \vee (x_1 = y_1 \wedge [x_2, \ldots, x_n] \leq_{\text{lex}} [y_2, \ldots, y_n])$.

We say X is lexicographically greater than Y (denoted as $X >_{\text{lex}} Y$) if $X \leq_{\text{lex}} Y$ does not hold. The rows (columns) of a 2D-matrix are lexicographically ordered if each row (column) is lexicographically smaller than the next (if any).

Since the init-block of a CA is fixed before search starts, lexicographically ordering the rows and columns of the matrix is a bit complex. We have to make sure that the lexicographic order does not contradict the preset cells.

Assume that the ith row of the matrix is represented as

$$R_i = [ce_{i,1}, ce_{i,2}, \ldots, ce_{i,k}],$$

and the jth column of the matrix is represented as

$$C_j = [ce_{1,j}, ce_{2,j}, \ldots, ce_{N,j}].$$

The constraints for breaking the row and column symmetries in CA are as follows:

1. Suppose R_i and R_{i+1} ($i > b$) are two adjacent rows outside the init-block, then $R_i \leq_{\text{lex}} R_{i+1}$.
2. Suppose C_j and C_{j+1} ($j > t$) are two adjacent columns outside the init-block, then $C_j \leq_{\text{lex}} C_{j+1}$.
3. Suppose R_{i_1} ($i_1 \leq b$) is a row inside the init-block, and R_{i_2} ($i_2 > b$) is a row outside the init-block. If $ce_{i_1,j} = ce_{i_2,j}$ for all $1 \leq j \leq t$, then $R_{i_1} \leq_{\text{lex}} R_{i_2}$.

These symmetry breaking constraints are checked on-the-fly during the search. For example, consider $X = [1, 2, 3, 4]$ and $Y = [2, 3, ?, ?]$ where a "?" represents an unassigned cell. It is obvious that $X \leq_{\text{lex}} Y$.

Example 6.3 Consider the three matrices in Fig. 6.3. Matrix A satisfies all constraints. In Matrix B, the fourth column is lexicographically greater than the fifth column, violating the second constraint, hence Matrix B will not be encountered during the search. In Matrix C, the first row is lexicographically greater than the fifth row, violating the third constraint, so Matrix C will be precluded too.

6.3.3.2 Symbol Symmetries

For an arbitrary column c in a CA, all the cells in c share the same domain $D_c = \{0, 1, 2, \ldots, d_c - 1\}$, and these $|D_c|$ symbols are interchangeable, thus results in $2^{|D_c|} - 1$ symmetries. In constraint programming, symbol symmetries are also called symmetries of indistinguishable values, which can be tackled by imposing value precedence [4, 9].

The LNH (Least Number Heuristic) [12, 13] is a fairly simple yet powerful strategy to impose value precedence in a dynamic way. It is based on the observation that, at the early stages of the search, many symbols (which have not yet been used in the search) are essentially the same. Thus when choosing a value for a new cell, there is no need to consider all the symbols in the domain. It makes no difference to assign the symbol e or e' to the cell, if neither of them has been previously assigned.

When processing column c, all the unused symbols are indistinguishable for the unassigned cells. We use a program variable mdn_c to represent the largest symbol that has been assigned in column c. Then the candidate domain for the next unassigned cell is $\{0, 1, 2, \ldots, mdn_c + 1\}$. In fact, $(mdn_c + 1)$ is a representative of all currently unused symbols. The variable mdn_c is initially set to -1 and dynamically updated during the search. Since in the first t columns, all symbols are covered by the init-block, LNH can only be applied to columns outside the init-block.

Acctually, the effect of LNH strategy is equivalent to imposing the following constraint:

$$ce_{i+1,c} \leq \text{MAX}\{ce_{1,c}, \ldots, ce_{i,c}\} + 1$$

where $t < c < k$ and $0 < i < N$.

6.3.4 Sub-combination Equalization Heuristic

Meagher and Stevens [6] conjectured that it is always possible to find a pairwise CA where the symbols in a row have balanced frequencies of occurrence (which is called Balanced Covering Array). Yan and Zhang [10] generalized the conjecture to CAs of strength t and proposed a hypothesis named sub-combination equalization. Based on observations, the hypothesis suggests that in any s ($s < t$) columns of the CA, each value combination appears *almost the same* number of times.

Definition 6.3 (*Sub-combination*) In a CA, a sub-combination of size s is a vector of s values from s columns, denoted as SC_C^V, where V is the vector of values and C

Fig. 6.4 The value vector's frequency

(a)

$$
\begin{array}{ccccc}
a & a & a & a & a \\
a & b & b & b & b \\
a & c & b & c & c \\
b & a & a & b & c \\
b & b & c & a & c \\
b & c & a & a & b \\
c & a & b & a & c \\
c & b & a & c & b \\
c & c & c & b & a \\
a & a & c & c & b \\
b & b & b & c & a \\
\end{array}
$$

(b)

The Value Vectors	$\langle a \rangle$	$\langle b \rangle$	$\langle c \rangle$
Column 1	4	4	3
Column 2	4	4	3
Column 3	4	4	3
Column 4	4	3	4
Column 5	3	4	4

the vector of columns. The frequency of the sub-combination SC_C^V, denoted by f_C^V, is the number of occurrences of the s-tuple V in the columns C.

Definition 6.4 (*Sub-Combination Equalization Heuristic (SCEH)*) In a CA of strength t, for any two sub-combinations $SC_C^{V_1}$ and $SC_C^{V_2}$ of size s ($0 < s < t$) from the same column set C, we have

$$|f_C^{V_1} - f_C^{V_2}| \le 1 \text{ for all } V_1 \ne V_2.$$

Example 6.4 A covering array $CA(11, 3^5, 2)$ is given in Table (A) of Fig. 6.4. Consider $s = 1$, then we have $\binom{k}{s} = 5$ column vectors and each has 3 value vectors. We can count the frequency of each column vector. The results are listed in Table (B). The frequencies of different vectors with the same column are almost the same.

As a mathematical hypothesis, SCEH has neither been proved nor disproved so far to our knowledge. Nevertheless, it can serve as an effective strategy to prune the search space. For almost all CA instances, the search time can be reduced with this strategy.

While processing a CA, one can specify the strategy level SL, so that the occurrences of all s-tuples ($0 < s \le$ SL) are restricted. Normally we would choose SL $= t - 1$. The strategy is applied to the backtracking search algorithm by checking the upper-bound UB and lower-bound LB of the frequency of each sub-combination. If an assignment $x = u$ causes the frequency of a sub-combination to go beyond the range [LB, UB], then the value u will be eliminated from the candidate set.

6.3.5 Combination of Strategies

In previous sections, we have introduced four strategies to reduce the search space, including: init-block, double lexicographic order, LNH and SCEH. Now an important question naturally arises: Can these strategies be used together during the search? The answer is positive. All of the four strategies can be combined to guide the searching without losing any non-isomorphic solution.

Since SCEH only restricts the frequency of each sub-combination, it has no influence on the other three strategies. Now we shall show that if a CA problem has a solution, then we can always find at least one matrix which contains an init-block and satisfies the constraints of double lexicographic order and LNH.

Suppose matrix A is a solution to a CA problem. First the rows are permuted so that in the first b rows, each value combination from the first t columns occurs exactly once, and these b rows are in non-descending lexicographic order (i.e., for all $0 < i < b$, $R_i \leq_{lex} R_{i+1}$). Now A is a CA with an init-block.

Then A is adjusted with the following four operations until all constraints of double lexicographic order and LNH are satisfied. All these operations preserve that A is a solution:

- Sort the columns outside the init-block so that the columns with the same level are in non-descending lexicographic order.
- Sort the rows outside the init-block in non-descending lexicographic order.
- If R_{i_1} ($i_1 \leq b$) and R_{i_2} ($i_2 > b$) share the same values in the first t cells, and $R_{i_1} >_{lex} R_{i_2}$, then swap R_{i_1} with R_{i_2}.
- In column C_j ($t < j \leq k$), denote the position of the first appearance of symbol s in C_j by FA_s. If there are two symbols s_1 and s_2 such that $s_1 < s_2$ and $FA_{s_1} > FA_{s_2}$, permute s_1 and s_2.

The order of these operations is not important.

If we regard the matrix A as a vector

$$A = [R_1, R_2, \ldots, R_N]$$
$$= [ce_{1,1}, \ldots, ce_{1_k}, \ldots, ce_{N,1}, \ldots, ce_{N,k}],$$

then each adjustment makes A lexicographically smaller. Since the number of matrices is bounded, the procedure will come to an end. Finally all the constraints are satisfied.

6.3.6 Constraint Propagation

When searching for a CA, the requirments of CA and the SCEH strategy can be directly utilized to detect inconsistency in constraint propagation. In order to reveal deeper inconsistency, we need more insight into the structure of the problem. Here we introduce a method based on graph theory.

Suppose we are processing column c. Consider a column combination C containing column c. We use X to represent the set of uncovered value combination of C and Y to denote the set of unassigned cells. We can build a bipartite graph G with X and Y as partite sets. The edges are defined as coverage relations: there is an edge

Fig. 6.5 Bipartite Graph
(Reprinted with permission
from [10]. Copyright 2008,
Elsevier Inc.)

between $x \in X$ and $y \in Y$ if the value combination x can be covered by assigning a value to cell y. The goal is to find a matching that each value combination is linked to a cell, i.e., a matching in G that saturates set X.

P. Hall proved the following classical theorem in 1935:

Theorem 6.1 *(Hall's Marriage Theorem) Let G be a bipartite graph with partite sets X and Y. There is a matching in G saturating X iff $|N(S)| \geq |S|$ for every $S \subseteq X$, where $N(S)$ is the neighbor set of S which is defined as the set of vertices adjacent to the vertices in S.*

X has $2^{|X|}$ subsets, of which we can only choose a few for pruning. For example, consider $S = X$, since $N(X) \leq |Y|$, if $|Y| < |X|$, there is a contradiction.

Example 6.5 Suppose we are processing a column c_3 of a CA, and c_3 has 3 cells unassigned. Assume $t = 3$. Consider 3 columns c_1, c_2 and c_3 ($c_1 < c_2 < c_3$). If there are three 3-tuple $X = \{\langle 2, 1, 0 \rangle, \langle 1, 1, 1 \rangle, \langle 1, 1, 0 \rangle\}$ to be covered, then we can draw the bipartite graph in Fig. 6.5, where "?" denotes the unassigned cells. Let Y be the neighbor set of X, obviously $|Y| = 2 < |X| = 3$, so we can not assign these 3 cells to cover X.

6.3.7 An Example of the Search Process

Figure 6.6 is an example of the search tree describing the process of constructing pairwise covering array CA($5, 2^4, 2$). The notation "?" represents an unassigned cell. We use the SCEH strategy level 1 (i.e., $SL = 1$).

The process begins at State I in which the cells of the init-block are assigned. Then we try to assign possible values to the cells of column 3. The cells $ce_{4,3}$ and $ce_{5,3}$ are assigned the value 1 according to SCEH at State II. At State III, two cells $ce_{5,1}$ and $ce_{5,2}$ are assigned consecutively to cover all the pairs of the first 3 columns. At State IV, $ce_{1,4}$ can only get the value 0 according to LNH. Then there are 2 candidate values 0 and 1 for $ce_{2,4}$. At State V, we assign 0 to $ce_{2,4}$. At the next State VI, if $ce_{3,4}$ is assigned 0, then according to Marriage Theorem, there are 3 uncovered combinations of column 3 and 4 (the set X), and only 2 unassigned cells left (the set Y), $|Y| < |X|$. We can not find a match and thus the value 0 for $ce_{3,4}$ is invalid. So we assign 1 to $ce_{3,4}$ and reach State VI. At the State VI, no candidate value for $ce_{4,4}$ is valid because of Marriage Theorem. So we backtrack to State IV. Similarly, we also find contradiction at State VIII. At last we find a covering array at State XI. If we need more than one covering arrays, we can backtrack and continue the search.

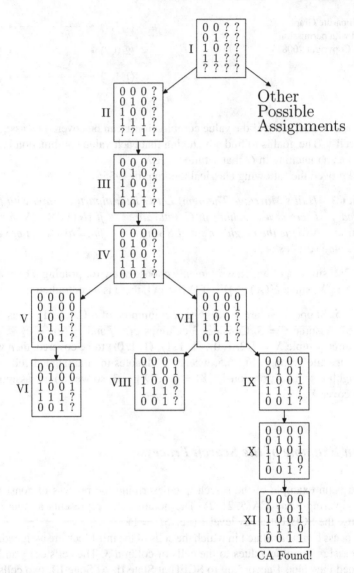

Fig. 6.6 Search Tree (Reprinted with permission from [10]. Copyright 2008, Elsevier Inc.)

6.3.8 A New Smaller CA

The CA finding tool EXACT [10] is based on the above strategies. The tool found the CA($24, 2^{12}, 4$) in Fig. 6.7 which is much smaller than previously known.

Fig. 6.7 A New
CA(24, 2^{12}, 4)

```
0 0 0 0 0 0 0 0 0 0 0 0
0 0 0 1 0 0 0 1 1 1 1 1
0 0 1 0 0 1 1 0 0 1 1 1
0 0 1 1 1 0 1 0 1 0 0 1
0 1 0 0 1 0 1 1 0 0 1 1
0 1 0 1 0 1 1 0 1 0 1 0
0 1 1 0 0 1 0 1 1 0 0 1
0 1 1 1 0 0 1 1 0 1 0 0
1 0 0 0 1 1 0 0 1 0 1 1
1 0 0 1 0 1 1 1 0 0 0 1
1 0 1 0 0 0 1 1 1 0 1 0
1 0 1 1 0 1 0 0 1 1 0 0
1 1 0 0 0 0 1 0 1 1 0 1
1 1 0 1 1 0 0 1 1 0 0 0
1 1 1 0 1 1 1 0 0 0 0 0
1 1 1 1 0 0 0 0 0 0 1 1
0 0 0 0 1 1 1 1 1 1 0 0
0 0 1 1 1 1 0 1 0 0 1 0
0 1 0 1 1 1 0 0 0 1 0 1
0 1 1 0 1 0 0 0 1 1 1 0
1 0 0 1 1 0 1 0 0 1 1 0
1 0 1 0 1 0 0 1 0 1 0 1
1 1 0 0 0 1 0 1 0 1 1 0
1 1 1 1 1 1 1 1 1 1 1 1
```

6.4 Backtracking Search for Generating Orthogonal Arrays

Since orthogonal arrays are a special kind of covering arrays, their generation techniques resemble a lot. However, if we encode the requirements on orthogonal arrays
using propositional formulas, there will be too many variables and clauses. Thus it
is not wise to transform the whole problem into the satisfiability problem and use a
SAT solver to find an orthogonal array.

In this section, we describe an approach proposed by Ma and Zhang [5]. A related
work is done by Schoen, Eendebak and Nguyen [8] who proposed an algorithm to
enumerate a minimum complete set of non-isomorphic orthogonal arrays. We shall
not describe that algorithm in detail, because in software testing, we usually need
just one orthogonal array.

6.4.1 Preprocessing

Assume that we are trying to find an $\mathrm{OA}(N, s_1 \cdot s_2 \ldots s_k, t)$, and the levels $s_1, s_2, \ldots, s_k$
are sorted in non-increasing order. Noticing the fact that in the first t columns all
combinations of $s_1, \ldots, s_t$ symbols should appear $\lambda = \frac{N}{s_1 \times \cdots \times s_t}$ times, we can generate an init-block directly by enumerating all possibilities, each of which λ times.

Fig. 6.8 OA(8, 2^4, 3)

$$
\begin{array}{ccc|c}
0 & 0 & 0 & 0 \\
0 & 0 & 1 & 1 \\
0 & 1 & 0 & 1 \\
0 & 1 & 1 & 0 \\
1 & 0 & 0 & 1 \\
1 & 0 & 1 & 0 \\
1 & 1 & 0 & 0 \\
1 & 1 & 1 & 1 \\
\end{array}
$$

For example, in Fig. 6.8, the 8×3 sub-array formed by the first three columns of OA(8, 2^4, 3) is an init-block.

6.4.2 The Basic Procedure

The approach is based on backtracking search, finding an OA *column by column*. It can be described as a recursive procedure in Algorithm 5.

Algorithm 5 Backtracking Search bool Col_Sch(j)

1: *cons* = Cons_Gen(j, *oaSol*);
2: **while TRUE do**
3:   *column* = Solve(*cons*);
4:   **if** *column* == *null* **then**
5:     return **FALSE**;
6:   **end if**
7:   Append(*column*, *oaSol*);
8:   **if** $j == k$ **or** Col_Sch($j + 1$) **then**
9:     return **TRUE**;
10:   **end if**
11:   Delete(*column*, *oaSol*);
12:   add Negate(*column*) to *cons*;
13: **end while**

Suppose we are processing the jth column and currently the obtained partial solution is denoted by *oaSol*. The function Cons_Gen(j) generates some constraints (denoted by *cons* in the pseudo-code) which are both necessary and sufficient for the jth column to satisfy the OA requirments. Then the function Solve(*cons*) tries to find a solution (denoted by *column*). If such a solution exists, it is added to *oaSol* by the function Append. When all the k columns are generated, an OA is obtained and the function returns TRUE. Otherwise, *oaSol* is not completed, and the procedure is executed recursively. If there is no solution to the $(j + 1)$th column, the algorithm backtracks to the jth column, deletes *column* from *oaSol*, and tries to generate another solution.

Fig. 6.9 Stack of p-sets

Column	P-set
1 2	$\{1, 2\}$ $\{3, 4\}$ $\{5, 6\}$ $\{7, 8\}$
1 3	$\{1, 3\}$ $\{2, 4\}$ $\{5, 7\}$ $\{6, 8\}$
2 3	$\{1, 5\}$ $\{2, 6\}$ $\{3, 7\}$ $\{4, 8\}$

Since the first t columns of *oaSol* are fixed in advance, the search procedure starts at column $t + 1$, and the first call of the recursive function is Col_Sch$(t + 1)$.

6.4.2.1 Constraint Generation

Assume that we have constructed m columns ($m \geq t$). How can we generate the constraints for column $m + 1$?

Theorem 6.2 *An* $OA(N, s_1 \cdot s_2 \ldots s_k, t)$ *is also an* $OA(N, s_1 \cdot s_2 \ldots s_k, t - 1)$.

If we extract $t - 1$ columns from an OA and partition the row indices by the row vectors in the subarray, we can get $s_{j_1} \times \cdots \times s_{j_{t-1}}$ mutually exclusive sets of equal size. Each set of the partition is called a **p-set** induced by the subarray.

Example 6.6 Consider the OA in Fig. 6.8. After the init-block is constructed, the p-sets are demonstrated in Fig. 6.9. For each 8×2 subarray in the init-block, there are four p-sets induced. More specifically, the p-set $\{1, 2\}$ is induced by the sub-array of column 1 and column 2 because row 1 and row 2 in the sub-array share the same row vector $\langle 0, 0 \rangle$. Similarly, if we examine the rest row vectors of column 1 and column 2, we would get the partition $\{\{1, 2\}, \{3, 4\}, \{5, 6\}, \{7, 8\}\}$.

Theorem 6.3 *An* $N \times (m + 1)$ *matrix is an* $OA(N, s_1 \cdot s_2 \ldots s_m \cdot s_{m+1}, t)$ **iff** *the matrix* A_m *formed by its first* m *columns is an* $OA(N, s_1 \cdot s_2 \ldots s_m, t)$, *and for each* $N \times (t - 1)$ *subarray of* A_m, *the* s_{m+1} *symbols in column* $m + 1$ *are equally distributed within the rows in each p-set induced by the subarray.*

According to Theorem 6.3, firstly we should calculate all p-sets induced by all $N \times (t - 1)$ sub-arrays from the first m columns, then the constraints for column $m + 1$ can be obtained directly.

After the constraints are obtained, they can be checked by an efficient SAT solver or pseudo-Boolean constraint solver [5].

6.4.3 Exploiting Symmetries

Like CAs, OAs also have row & column symmetries as well as symbol symmetries. For example, Fig. 6.10 illustrates two isomorphic OAs.

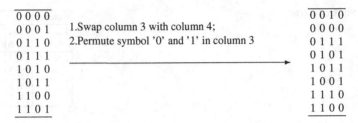

Fig. 6.10 Two isomorphic instances of OA($8, 2^4, 2$)

To eliminate row & column symmetries, lexicographic order is imposed along both rows and columns in an OA. Note that for an OA with mixed factor levels, column lexicographic order is only imposed on the columns with the same levels. To break symbol symmetries, the LNH strategy can be applied to each column outside the init-block.

Example 6.7 Figure 6.11 demonstrates three solutions of OA($12, 3 \cdot 2^4, 2$). A and B are lexicographically ordered along both the rows and the columns. Matrix C does not satisfy the lexicographic order since the third column $\not\leq_{lex}$ the fourth column, thus C would not be encountered in the search process.

Although being quite effective, lexicographic order and LNH are far from enough to eliminate all symmetries in OAs. There is a class of symmetries arising from the automorphisms of init-block. If we permute the symbols in the init-block, reconstruct the init-block by swapping rows, we can always get another OA by performing some other isomorphic operations outside the init-block. The newly obtained array has the same init-block, satisfies all constraints of lexicographic order and LNH, hence would also be encountered during the search. This transformation procedure is called *init-block reconstruction*.

We say two OAs are *symmetric with respect to (w.r.t.) an init-block reconstruction* if one can be obtained from the other through this reconstruction.

Fig. 6.11 Three Instances of OA($12, 3 \cdot 2^4, 2$)

(a)
```
0 0|0 0 0
0 0|1 1 1
0 1|0 1 1
0 1|1 0 0
1 0|0 0 1
1 0|0 1 0
1 1|1 0 0
1 1|1 1 1
2 0|1 0 1
2 0|1 1 0
2 1|0 0 1
2 1|0 1 0
```

(b)
```
0 0|0 0 0
0 0|0 1 1
0 1|1 0 1
0 1|1 1 0
1 0|0 0 1
1 0|1 1 0
1 1|0 1 0
1 1|1 0 1
2 0|1 0 0
2 0|1 1 1
2 1|0 0 0
2 1|0 1 1
```

(c)
```
0 0|0 0 0
0 0|1 0 1
0 1|0 1 0
0 1|1 1 1
1 0|0 1 1
1 0|1 1 0
1 1|0 0 0
1 1|1 0 1
2 0|0 1 1
2 0|1 0 0
2 1|0 0 1
2 1|1 1 0
```

Example 6.8 In Fig. 6.11, we can obtain Matrix B from A by performing the following init-block reconstruction:
1. Permute symbol '0' and '1' in the first column.
2. Swap rows 1, 2, 3, 4 with rows 5, 6, 7, 8 respectively.
3. Permute symbol '0' and '1' in the last column.

After step 2, the init-block is unchanged. However, the last column of Matrix A is converted to $\langle 100101101010 \rangle$, contradicting Formula (1) which specifies that symbol '0' must precede '1'. Step 3 is then performed to adjust the matrix and we get Matrix B, which satisfies all symmetry breaking constraints. Therefore, A and B are symmetric w.r.t. the init-block reconstruction.

For an $OA(N, s_1 \cdot s_2 \ldots s_k, t)$, there are $s_1! \times s_2! \ldots s_t! - 1$ symmetries caused by init-block reconstructions except for the identity mapping. To break all these symmetries, it can be costly, because there are too many swappings and permutations to perform.

Another technique called **Filter** [5] is also useful to break such symmetries. The basic idea is to add additional constraints to the column adjoining the init-block, i.e. the $(t + 1)$th column. Once a symmetry caused by init-block reconstruction has been broken in the $(t + 1)$th column, it is prevented from spreading to the following columns. It is like setting a filter beyond the init-block, hence the name filter. Filter can not break all symmetries w.r.t. init-block reconstruction, but it can eliminate enough isomorphisms, yet the extra cost is negligible.

References

1. Cohen, M.B., Dwyer, M.B., Shi, J.: Constructing interaction test suites for highly-configurable systems in the presence of constraints: a greedy approach. IEEE Trans. Softw. Eng. **34**(5), 633–650 (2008)
2. Flener, P., Frisch, A.M., Hnich, B., Kiziltan, Z., Miguel, I., Pearson, J., Walsh, T.: Breaking row and column symmetries in matrix models. In: Proceedings of 8th International Conference on Principles and Practice of Constraint Programming, pp. 462–477 (2002)
3. Hnich, B., Prestwich, S.D., Selensky, E., Smith, B.M.: Constraint models for the covering test problem. Constraints **11**(2–3), 199–219 (2006)
4. Law, Y.C., Lee, J.H.: Symmetry breaking constraints for value symmetries in constraint satisfaction. Constraints **11**(2–3), 221–267 (2006)
5. Ma, F., Zhang, J.: Finding orthogonal arrays using satisfiability checkers and symmetry breaking constraints. In: Proceedings of the 10th Pacific Rim International Conference on Artificial Intelligence (PRICAI), pp. 247–259 (2008)
6. Meagher, K., Stevens, B.: Covering arrays on graphs. J. Comb. Theory Ser. B **95**(1), 134–151 (2005)
7. Rossi, F., van Beek, P., Walsh, T. (eds.): Handbook of Constraint Programming. Elsevier, Amsterdam (2006)
8. Schoen, E.D., Eendebak, P.T., Nguyen, M.V.M.: Complete enumeration of pure-level and mixed-level orthogonal arrays. J. Comb. Des. **18**(2), 123–140 (2010)
9. Walsh, T.: Symmetry breaking using value precedence. In: Proceedings ECAI-06, pp. 168–172 (2006)

10. Yan, J., Zhang, J.: A backtracking search tool for constructing combinatorial test suites. J. Syst. Softw. **81**(10), 1681–1693 (2008)
11. Yu, L., Lei, Y., Nourozborazjany, M., Kacker, R. N., Kuhn, D. R.: An efficient algorithm for constraint handling in combinatorial test generation. In: Proceedings of the Sixth International Conference on Software Testing, Verification and Validation (ICST'13), pp. 242–251 (2013)
12. Zhang, J.: Automatic symmetry breaking method combined with SAT. In: Proceedings of the ACM Symposium on Applied Computing (SAC'01), pp. 17–21 (2001)
13. Zhang, J., Zhang, H.: SEM: a system for enumerating models. In: Proceedings of IJCAI-95, pp. 298–303 (1995)

Chapter 7
Tools and Benchmarks

Abstract This chapter lists some test generation tools (mostly, tools for finding covering arrays). The chapter also gives some pointers to benchmarks, so that the reader can evaluate and select tools.

7.1 Test Input Generation Tools

During the past twenty years or so, many automatic tools have been developed, which can help us design combinatorial test suites. Some of the tools have been mentioned in the previous chapters of this book.

In this section, we briefly describe some of the automatic test generation tools for combinatorial testing.

Jacek Czerwonka maintains a list of tools for combinatorial testing (in particular, pairwise testing). See http://www.pairwise.org/. Some of the tools are commercial; some are free.

- ACTS (http://csrc.nist.gov/groups/SNS/acts/index.html) [15] is a powerful combinatorial test generation tool developed by Rick Kuhn and Raghu Kacker (US National Institute of Standards and Technology), Jeff Yu Lei (University of Texas at Arlington), and others. It implements the IPO algorithms (including IPOG, IPOG-D, IPOG-F, IPOG-F2), as described in Chap. 4. ACTS can be used to generate t-way test suites ($1 \leq t \leq 6$). It can construct mixed strength CAs, and it allows the user to specify constraints.
- AETG is one of the first test data generation systems for combinatorial testing. Its main algorithm has been described previously, in Chap. 3. The system is now available as a web service (http://aetgweb.argreenhouse.com/). The generated test cases can be used for manual testing or to drive some other automated test tools.
- Allpairs (http://www.satisfice.com/tools.shtml) by James Bach.
- BOAS (http://lcs.ios.ac.cn/~zj/ct.html) focuses on the generation of orthogonal arrays (OAs). The basic algorithm is backtracking search, described in Chap. 6.

© The Author(s) 2014

J. Zhang et al., *Automatic Generation of Combinatorial Test Data*,

SpringerBriefs in Computer Science, DOI 10.1007/978-3-662-43429-1_7

- CASA (http://cse.unl.edu/~citportal/) is a tool for generating Covering Arrays by Simulated Annealing [7].
- Cascade (http://lcs.ios.ac.cn/~zj/ct.html) is based on the one-test-at-a-time strategy, and it uses pseudo-Boolean optimization (PBO) techniques. See Sect. 3.4. The tool can search for variable-strength covering arrays with constraints.
- CitLab (http://code.google.com/a/eclipselabs.org/p/citlab/) [2] is a Laboratory for Combinatorial Interaction Testing [6]. It provides an abstract language for combinatorial testing problems, so as to promote the interoperability among test data generation tools, and to make the exchange of models and data more convenient.
- EXACT [14] (http://lcs.ios.ac.cn/~zj/ct.html) tries to find a small covering array by backtracking search. For more details, see Chap. 6. EXACT has been used to find several arrays, which are smaller than what people knew.
- Jenny (http://burtleburtle.net/bob/math/jenny.html) is an open-source tool for pairwise testing, implemented in C.
- PICT (Pairwise independent combinatorial testing) is a widely used test generator, developed by Jacek Czerwonka (Microsoft Corp.) [5].
- QICT [12] is an open-source pairwise test data generator, implemented in C#.
- Testcover.com (http://testcover.com/) provides a test case generator for pairwise testing. It is also available as a web service.
- Test Vector Generator (TVG) is available at http://sourceforge.net/projects/tvg.

When we need to use combinatorial testing in a project, which tools shall we choose? How do the above tools compare with each other?

Researchers have some general observations. For instance, greedy techniques tend to be faster, while meta-heuristic search techniques can often find smaller test suites. There are also some empirical evaluations of the different approaches, e.g., Grindal et al. [8]. However, as more methods and tools are being developed, and with the advance in hardware performance, the situation may change. Another factor to consider is that some algorithms involve a certain kind of randomness. If you execute the tools several times, you may get several different results.

Thus, it seems to be difficult to give a simple answer saying that one tool is better than another. We encourage the readers to try the available tools on well-known benchmarks and on your own applications.

7.2 Applications and Benchmarks

Combinatorial testing is closely related to combinatorics, which is a branch of mathematics. Mathematicians have published many papers on combinatorial designs. But there are still many open questions. On the other hand, as a software/system testing technique, combinatorial testing has been used in practice by quite some researchers and engineers.

Both mathematicians and software engineering researchers have collected some benchmarks, which have helped to advance the state of the art in the generation of small covering arrays. For example,

- Charlie Colbourn maintains a collection of covering arrays of strength $t = 2, 3, 4, 5, 6$. The sizes are the smallest; but the tables are not shown. See http://www.public.asu.edu/~ccolbou/src/tabby/catable.html.
- Sloane maintains a Library of Orthogonal Arrays at http://neilsloane.com/oadir/.
- US National Institute of Standards and Technology (NIST) also publishes some covering arrays (http://math.nist.gov/coveringarrays/). The tables are given explicitly; but they are not necessarily optimal in terms of size.
- IBM Haifa Research Lab [13] collected several testing problems from different domains, such as telecommunication, health care, storage, and banking, and for testing different aspects of the system, such as data manipulation and integrity, protocol validation, and user interface testing. The benchmarks are available at http://researcher.watson.ibm.com/researcher/files/il-ITAIS/ctdBenchmarks.zip.

In Chap. 1, we described some applications of combinatorial testing. In this section, we will present a few more applications and benchmarks.

Ahmed and Zamli [1] selected `flex` as a case study. It is a lexical analyzer, which was obtained from the Software-artifact Infrastructure Repository (SIR— http://sir.unl.edu/portal/index.php). Ahmed and Zamli applied variable strength covering arrays to the testing of `flex`. They considered 11 parameters, 5 of which were exercised with higher interaction strength.

Cohen, Dwyer, and Shi [3, 4] studied several nontrivial highly configurable software systems: SPIN (model checker and simulator), the GCC compiler, Apache HTTP server and Bugzilla.

- SPIN [9, 10] can be used as a verifier or as a simulator. The simulator has 18 factors: 13 binary options, and 5 options each with 4 different values. The verifier's configuration model consists of 55 factors: ($2^{42} \cdot 3^2 \cdot 4^{11}$). This model includes a total of 49 constraints.
- After analyzing the documentation of the GCC Compiler (version 4.1),[1] Cohen, Dwyer, and Shi constructed three different configuration models: (1) a comprehensive model accounting for all of the GCC options, which has 1462 factors and 406 constraints; (2) a model that eliminates some machine-specific options, which has 702 factors and 213 constraints; and (3) a model that focuses on the machine-independent optimizer of GCC, which has 199 factors and 40 constraints.
- Apache HTTP server is a popular web server. Cohen, Dwyer, and Shi studied the documentation for the server (version 2.2) and established a combinatorial testing model, which has 172 factors: ($2^{158} \cdot 3^8 \cdot 4^4 \cdot 5^1 \cdot 6^1$). They also identified seven constraints.

---

[1] http://gcc.gnu.org/onlinedocs/gcc-4.1.1/gcc/.

- Bugzilla (http://www.bugzilla.org/) is an open source defect tracking system. Cohen, Dwyer, and Shi examined part of the user manual and established a combinatorial testing model, which has 52 factors: $(2^{49} \cdot 3^1 \cdot 4^2)$. There are five constraints in the documentation.

In the book [11] (Chap. 5, "Test Parameter Analysis", by Eduardo Miranda), several case studies are given:

- a flexible manufacturing system, which involves a conveyor, a camera, a machine vision system ,and a robotic arm subsystem. It picks up metal sheets of different shapes, colors, and sizes, and puts them into appropriate bins.
- an audio amplifier, which has two input jacks, two volume controls, two toggle switches, and one three-way selection switch.
- a password diagnoser for an online banking system, which verifies that the user passwords conform to good security practices. For example, the length of a password should be at least eight (characters); a password should have at least one uppercase character, one numeric character, and one special character.

References

1. Ahmed, B.S., Zamli, K.Z.: A variable strength interaction test suites generation strategy using particle swarm optimization. J. Syst. Softw. **84**(12), 2171–2185 (2011)
2. Calvagna, A., Gargantini, A., Vavassori, P.: Combinatorial interaction testing with CitLab. In: Proceedings of IEEE Sixth International Conference on Software Testing, Verification and Validation (ICST), pp. 376–382 (2013)
3. Cohen, M.B., Dwyer, M.B., Shi, J.: Interaction testing of highly-configurable systems in the presence of constraints. In: Proceedings of the International Symposium on Software Testing and Analysis (ISSTA), pp. 129–139 (2007)
4. Cohen, M.B., Dwyer, M.B., Shi, J.: Constructing interaction test suites for highly-configurable systems in the presence of constraints: a greedy approach. IEEE Trans. Software Eng. **34**(5), 633–650 (2008)
5. Czerwonka, J.: Pairwise testing in real world: practical extensions to test case generator, In: Proceedings of 24th Pacific Northwest Software Quality Conference, pp. 419–430 (2006)
6. Gargantini, A., Vavassori, P.: CITLAB: A laboratory for combinatorial interaction testing. Workshop on Combinatorial Testing (CT). In: Proceedings of the International Conference on Software Testing, Verification and Validation (ICST 2012), pp. 559–568 (2012)
7. Garvin, B.J., Cohen, M.B., Dwyer, M.B.: An improved meta-heuristic search for constrained interaction testing. Proc. of the International Symposium on Search-Based Software Engineering (SSBSE 2009), pp. 13–22 (2009)
8. Grindal, M., Lindström, B., Offutt, J., Andler, S.F.: An evaluation of combination strategies for test case selection. Empir. Software Eng. **11**(4), 583–611 (2006)
9. Holzmann, G.J.: The model checker SPIN. IEEE Trans. Software Eng. **23**(5), 279–295 (1997)
10. Holzmann, G.J.: Spin Man pages. http://spinroot.com/spin/Man/index.html
11. Kuhn, D.R., Kacker, R., Lei, Y.: Introduction to Combinatorial Testing. Chapman & Hall / CRC, London (2013)
12. McCaffrey, J.: Pairwise testing with QICT. MSDN magazine. http://msdn.microsoft.com/en-us/magazine/ee819137.aspx. (2009)

13. Segall, I., Tzoref-Brill, R., Farchi, E.: Using binary decision diagrams for combinatorial test design. In: Proceedings of the International Symposium on Software Testing and Analysis (ISSTA), pp. 254–264 (2011)
14. Yan, J., Zhang, J.: A backtracking search tool for constructing combinatorial test suites. J. Syst. Softw. **81**(10), 1681–1693 (2008)
15. Yu, L., Lei, Y., Kacker, R., Kuhn, D.R.: ACTS: a combinatorial test generation tool. In: Proceedings of IEEE Sixth International Conference on Software Testing, Verification and Validation (ICST), pp. 370–375 (2013)

Chapter 8
Other Related Topics

Abstract This chapter briefly discusses some other topics that are related to test data generation. They include, among others, how to construct a combinatorial testing model for the software under test, how to select a subset of test cases from an existing test suite, and how to generate special-purpose test cases in the process of debugging.

8.1 Problems Related to Test Input Generation

In the previous chapters, we described many techniques and tools for generating the input data for combinatorial testing (CT). We have taken a simplistic view of the problem: Given a model (consisting of parameters, values, constraints, etc.), find a small (or perhaps the smallest) test suite satisfying the constraints and achieving the required coverage.

In practice, test input generation is a part of the software development process. There are several other related activities in the process, like test case execution and debugging. See Fig. 8.1.

To apply combinatorial testing to the system under test (SUT), we usually need a model (unless we already have some test cases). Given that model, we can use various techniques and tools (as described in the previous chapters) to obtain a suitable test suite. Then, we test the system using the test suite. Ideally, all the test cases pass. But it may happen that a few test cases fail, which means the system has some defect(s). Then, we should identify a small set of parameters, which leads to the failure. Such a combination of parameters may help us find the bugs in the system. After the bugs are fixed or the developers change the system due to new requirements, we should test the system again. This is an iterative process.

Given the above *global* view of the software development process, we may ask some questions about testing. For example, can we reuse some existing test cases? If not, we need to generate a test suite. But where does the model come from? Is it always good to generate a very small set of test cases? Do we need to put the test cases in certain order?

© The Author(s) 2014

J. Zhang et al., *Automatic Generation of Combinatorial Test Data*,
SpringerBriefs in Computer Science, DOI 10.1007/978-3-662-43429-1_8

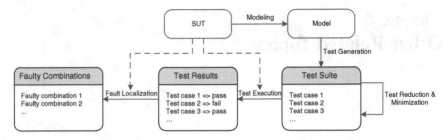

Fig. 8.1   Related activities

8.2 Modeling

To use the test data generation methods described in the previous chapters, we should have a suitable model of the software/system, including the constraints. Currently, in most cases, the model is constructed by the test designers manually. The model can be built from the SUT itself, or from the system requirements or design specification.

We assume that there are a number of parameters in the CT model, each of which can take values from a discrete domain, which is not too large. However, in real software systems, many input variables have very large ranges of values. Thus, to use combinatorial testing techniques, we have to build an abstract model using various other methods, e.g., that based on equivalence classes.

Ghandehari et al. [1] described a three-step approach to apply CT to real code. First, an abstract model is created, which consists of abstract parameters and values. Then, a combinatorial test suite is generated from the abstract model. Finally, concrete tests are derived from the abstract tests. These concrete tests are used to perform the actual testing.

They used the Siemens programs to illustrate their approach. These programs were introduced to the software engineering community by several staff members of the Siemens corporation [2], and became famous benchmarks in software testing and analysis. The programs are relatively small in terms of lines of code, and they have a small number of input parameters. But the input spaces are quite complex. For example, one input parameter of the program *replace* is a regular expression, and we need to deal with some metacharacters in it. This program has less than 1,000 lines of code and just 3 input parameters. But its abstract model for combinatorial testing [1] contains 20 abstract parameters and 36 constraints.

Instead of analyzing the input parameters of the code, we may also obtain a CT model from the software design specification. Recently, Satish et al. [3] presented an approach to extracting key information (such as parameters and values) of the CT model from UML sequence diagrams.

8.3 Test Case Prioritization and Test Suite Reduction

When the strength t is high, a combinatorial test suite achieving t-way coverage can be very large. The tester may not be able to run the entire test suite, due to time or budget constraints. Then, it is essential to *prioritize* the test cases [4].

To address the problem, Bryce and Colbourn [5] proposed an algorithm, which tries to maximize the number of t-tuples covered by the earlier tests so that if a tester only runs a partial test suite, he/she will be able to test as many t-tuples as possible. This work is extended in [6]. The authors note that "minimum test suite size and lowest expected time to fault detection are incompatible objectives". They differentiate between two goals, which are not so compatible: constructing a test suite of the smallest size versus covering as many t-way interactions as possible in the earliest tests. They developed a hybrid approach to test generation so as to reduce the expected time to fault detection. In this approach, a fast but effective greedy method is adopted to produce an initial candidate, and then it is modified repeatedly to improve the coverage. The test suite is constructed one test at a time while attempting to maximize the number of t-way interactions covered by the earliest tests.

Normally, a tester executes the whole test suite in sequence. But sometimes we need to reorder the test cases for some purposes. For example, some time and effort is needed when the tester changes from one test case to another. This is called the *switching cost*. Recently, several authors [7, 8] proposed methods for minimizing the switching cost while generating the covering array. On the other hand, Wu et al. [9] presented algorithms for prioritizing a given test suite, so as to minimize the total switching cost.

For some applications, using a complete new test suite might be too expensive, because the cost of implementing a test case is high (e.g., when it requires a large amount of specific data). To alleviate this problem, Blue et al. [10] proposed a method called Interaction-based Test-Suite Minimization (ITSM), to obtain a desired test suite from existing test cases. It is applicable when the company already has many test cases, which are quite extensive and representative. Instead of constructing a new test suite from scratch, ITSM reduces an existing test suite, while preserving its interaction coverage.

To use ITSM, one still needs to specify the parameters of the SUT and their values, but it is not necessary to define the constraints. Of course, there should also be a test suite. ITSM tries to select a subset of the test suite that preserves its t-combination coverage. ITSM does not guarantee full interaction coverage. The quality of the result depends on the existing test suite. But it may save a lot of effort when applicable. It can still be considered as a test generation method.

8.4 Fault Localization

Testing and debugging are different activities in the software development process, but they are closely related. After several bugs (defects) have been detected during testing, they need to be fixed. This bug-fixing process is called debugging. Typically, software debugging has two steps: the first step is to find the location of the cause of the bug, which is called *fault localization*, and the second step is to modify the code to eliminate the bug, which is called *bug fixing*. Fault localization often costs much time during debugging.

Unlike traditional fault localization, which targets at locating failure causes in the source code, fault localization based on combinatorial testing aims at locating failure-causing parameter combinations (also called *faulty combinations*). Since CT is a black-box testing technique, we usually have the SUT model only, without the source code. But finding the faulty combinations can still provide useful information for locating the root cause inside the SUT.

Colbourn and McClary, as well as Martínez et al. proposed to construct special-purpose covering arrays (such as *error locating arrays* [11, 12], *locating and detecting arrays* [13]) to identify the faulty combinations according to the test execution results. These covering arrays have fault localization abilities themselves, so that we can deduce the possible faulty combinations from the execution results. However, such arrays are often quite large.

An effective class of methods for fault localization uses adaptive testing, which generates and executes a set of additional test cases to provide more information for locating the fault, where the generation of additional test cases is based on the execution results of the previous test cases. Some algorithms start by analyzing the execution results of the test suite, and maintain a set of suspicious parameter combinations [14–16]. Then, they iteratively generate and execute several sets of additional test cases to narrow down the set of suspicious parameter combinations, until the faulty combinations are finally located. Some other algorithms start from a single failing test case, and then generate and execute some additional test cases to locate the faulty combinations in the failing test case [17–20]. The key observation of these algorithms is that if test case T_1 fails and T_2 passes, the non-overlapping part of T_1 and T_2 must contain some parameters of some faulty combination; and if both T_1 and T_2 fail, the overlapping part of T_1 and T_2 must contain a faulty combination.

Another category of CT-based fault localization methods is to analyze the execution results of the covering array. Intuitively, if most of the test cases containing a certain combination fail, this combination will be very likely to cause the failure. Classification tree [21] or SAT solving/optimization techniques [22] can be applied to find these combinations, given a set of execution results. Experimental results show that, when the size of the test suite is small, many possible faulty combinations will be found, and it is difficult to pinpoint the cause of the defects. The smallest test suite is not always the best for locating faults. Thus, it is necessary "to make a balance between reducing the size of the test suite and increasing the ability of fault localization" [22].

References

1. Ghandehari, L.S.G., Bourazjany, M.N., Lei, Y., Kacker, R.N., Kuhn, D.R.: Applying combinatorial testing to the Siemens suite. In: IEEE International Conference on Software Testing, Verification and Validation Workshops (ICSTW), pp. 362–371 (2013)
2. Hutchins, M., Foster, H., Goradia, T., Ostrand, T.J.: Experiments on the effectiveness of dataflow- and controlflow-based test adequacy criteria. In: Proceedings of the 16th International Conference on Software Engineering (ICSE), pp. 191–200 (1994)
3. Satish, P., Paul, A., Rangarajan, K.: Extracting the combinatorial test parameters and values from UML sequence diagrams. In: Proceedings of the IEEE International Conference on Software Testing, Verification and Validation Workshops (ICSTW'14), pp. 88–97 (2014)
4. Bryce, R.C., Colbourn, C.J.: Prioritized interaction testing for pair-wise coverage with seeding and constraints. Inf. Softw. Technol. **48**(10), 960–970 (2006)
5. Bryce, R.C., Colbourn, C.J.: One-test-at-a-time heuristic search for interaction test suites. In: Proceedings of 9th Conference on Genetic and Evolutionary Computation (GECCO), pp. 1082–1089 (2007)
6. Bryce, R.C., Colbourn, C.J.: Expected time to detection of interaction faults. J. Comb. Math. Comb. Comput., to appear (2013)
7. Kimoto, S., Tsuchiya, T., Kikuno, T.: Pairwise testing in the presence of configuration change cost. In: Proc. of the 2nd International Conference on Secure System Integration and Reliability Improvement (SSIRI), pp. 32–38 (2008)
8. Srikanth, H., Cohen, M.B., Qu, X.: Reducing field failures in system configurable software: Cost-based prioritization. In: Proceedings of the International Symposium on Software Reliability Engineering (ISSRE), pp. 61–70 (2009)
9. Wu, H., Nie, C., Kuo, F.-C.: Test suite prioritization by switching cost. In: Proceeidngs of the IEEE International Conference on Software Testing, Verification and Validation Workshops (ICSTW'14), pp. 133–142 (2014)
10. Blue, D., Segall, I., Tzoref-Brill, R., Zlotnick, A.: Interaction-based test-suite minimization. In: Proceedings of the 35th International Conference on Software Engineering (ICSE), pp. 182–191 (2013)
11. Martínez, C., Moura, L., Panario, D., Stevens, B.: Algorithms to locate errors using covering arrays. In: Proceedings of the 8th Latin American Symposium on Theoretical Informatics (LATIN), Springer Lecture Notes in Computer Science, vol. 4957, pp. 504–519 (2008)
12. Martínez, C., Moura, L., Panario, D., Stevens, B.: Locating errors using ELAs, covering arrays, and adaptive testing algorithms. SIAM J. Discrete Math. **23**(4), 1776–1799 (2009)
13. Colbourn, C.J., McClary, D.W.: Locating and detecting arrays for interaction faults. J. Comb. Optim. **15**(1), 17–48 (2008)
14. Ghandehari, L.S.G., Lei, Y., Xie, T., Kuhn, R., Kacker, R.: Identifying failure-inducing combinations in a combinatorial test set. In: IEEE International Conference on Software Testing, Verification and Validation (ICST), pp. 370–379 (2012)
15. Shakya, K., Xie, T., Li, N., Lei, Y., Kacker, R., Kuhn, R.: Isolating failure-inducing combinations in combinatorial testing using test augmentation and classification. In: Proceedings of the Fifth International Conference on Software Testing, Verification and Validation (ICST), pp. 620–623 (2012)
16. Wang, Z., Xu, B., Chen, L., Xu, L.: Adaptive interaction fault location based on combinatorial testing. In: Proceedings of the 10th International Conference on Quality Software (QSIC), pp. 495–502 (2010)
17. Niu, X., Nie, C., Lei, Y., Chan, A.T.: Identifying failure-inducing combinations using tuple relationship. In: IEEE International Conference on Software Testing, Verification and Validation Workshops (ICSTW), pp. 271–280 (2013)
18. Shi, L., Nie, C., Xu, B.: A software debugging method based on pairwise testing. In: Proceedings of the 5th International Conference on Computational Science (ICCS), Part III, Springer Lecture Notes in Computer Science, vol. 3516, pp. 1088–1091 (2005)

19. Zeller, A., Hildebrandt, R.: Simplifying and isolating failure-inducing input. IEEE Trans. Software Eng. **28**(2), 183–200 (2002)
20. Zhang, Z., Zhang, J.: Characterizing failure-causing parameter interactions by adaptive testing. In: Proceedings of the International Symposium on Software Testing and Analysis (ISSTA), pp. 331–341 (2011)
21. Yilmaz, C., Cohen, M.B., Porter, A.A.: Covering arrays for efficient fault characterization in complex configuration spaces. IEEE Trans. Software Eng. **32**(1), 20–34 (2006)
22. Zhang, J., Ma, F., Zhang, Z.: Faulty interaction identification via constraint solving and optimization. In: Proceedings of the 15th International Conference on Theory and Applications of Satisfiability Testing (SAT), Springer Lecture Notes in Computer Science, vol. 7317, pp. 186–199 (2012)